This is the definitive guide for anyone with the responsibility to select, juggle best support their communities of practice. An instant CoP classic.

Shawn Callahan, Anecdote Pty., http://www.anecdote.com

In the business world where plenty of focus has been placed around the topic of communities over the last couple of years, we may have reached that point in time where it's becoming yet another *buzzword*. Everything out there, every single grouping you can think of, seems to be flagged as *community* nowadays. But what **is really** a community? How do you get to nurture communities in a business environment? How do you manage to successfully build, sustain and help mature new, or already existing, communities? What community tools are available out there for them? Can new, emerging technologies from the Social Computing era help develop healthier communities in such a distributed world as today's?

Whether you are just getting involved with communities, whether you have been building communities for a while already, or whether you think there is nothing more to learn around the topic of communities in a business environment, "Digital habitats: stewarding technology for communities" by Etienne Wenger, Nancy White, and John Smith is just the right book for you to get answers to some of the questions mentioned above and many more!

In this particular, very well written, enlightening as well as educational, book you would be able to get plenty of hints and tips on how to get your hands around not only building successful communities, but also explore the mutual influence that technology has always been having in them. If there would be a single book that you would want to read this year around the topic of communities and the reciprocal influence of technology in them, including the fascinating role of social software and other community related tools, "Digital habitats: stewarding technology for communities" by Wenger, White, and Smith would be it. Without a doubt!

Luis Suarez, http://elsua.net - http://www.ittoolbox.com/profiles/elesar

Communities of practice have long been valuable to businesses by helping employees to share knowledge, help one another, and reuse proven practices. In Technology Stewardship for Communities, Etienne Wenger, Nancy White, and John D. Smith introduce technology stewarding, a new kind of community leadership. They help community leaders choose the right technology strategy, including using existing tools, free tools, enterprise platforms, commercial platforms, custom applications, open-source software, and elements patched together.

"Etienne, Nancy, and John are respected thought leaders, authors, bloggers, and community activists, particularly in CPsquare and com-prac. They practice what they preach, using communities to help learn about and improve communities of practice as a business discipline. This book provides a practical compilation of what they have learned, and it should be very useful to both new and experienced community leaders and practitioners.

Stan Garfield, knowledge manager and author, http://stangarfield.googlepages.com/

What a wonderful book! Although the audience focus is on those who need to understand, select, try out, introduce, troubleshoot, and enhance the use of web-based tools for communities of practice, this long-needed book provides working models, insights, and templates that can be applied to social networks of any type. Working with people and knowledge may be 90% people and only 10% technology (as many have said), but it is really important to get that 10% right: everything must work in the context of the current community. This book is the practitioner's sourcebook and is the much-needed companion to the COP's touchstone, Cultivating Communities of Practice. Only Etienne, John, and Nancy could have written this book, and I am so glad that they have.

Patti Anklam, author of Net Work: A Practical Guide to Creating and Sustaining Networks at Work and in the World, http://www.pattianklam.com

A practical guide for community facilitators who need to make technology choices--it helps identify the activities relevant to your community, the tools that might help, and the strategies to obtain them. It also provides helpful tips for making the most of technologies once they're up and running and a glimpse into the future of community technologies.

Eric Matson, Manager of Knowledge and Communities, Strategy Practice, McKinsey & Company

Digital habitats: stewarding technology for communities" is truly remarkable. For technology stewards at all levels of practice, it's inspiring, eminently practical, grounded in hard-won experience, and full of tips and lessons learned. Wenger, White, and Smith provide frameworks for thinking systemically about the wealth of technology tools available, how they might work together in practice, and how stewards find, choose, and use them to support diverse communities. Their focus is on the evolving interplay of community and technology -- and how each continually shapes the other. Whether you're a brand new technology steward, a "just do it-er" looking for concrete, practical information to use today, an "attentive practitioner" who wants to reflect on, develop, and deepen your stewarding practice; or a "deep diver" looking for insights about history, theory, and the future of technology for communities; this is the book for you.

Peter and Trudy Johnson-Lenz, groupware and virtual community pioneers, http://www.johnson-lenz.com

It's a wonderful book...
Many books are interesting, fewer are useful on the ground, but a book that combines deep, thoughtful insight with practical advice and guidance is a rarity. This book is one of those rarities. It is an incredibly thoughtful and useful book written by a tight team at the cutting edge of their practice. It will define the way that communities and networks look at - and take a more informed responsibility for - their technology platforms and tools. Knowledge managers have often been placed into a false polarisation between "people-oriented" initiatives (human and naturalistic) and "technology-oriented" initiatives (logical and constrained). In this book we start to recover a vision of technology as it was originally intended to be - an instrument of human action - and learn the principles for how to steward it wisely for collective learning, collaboration and growth.

Patrick Lambe, author of www.organisingknowledge.com, www.greenchameleon.com and www.straitsknowledge.com

Only a few years ago, we used to routinely talk about technology merely "enabling" Communities of Practice. This simplistic approach is now hopefully merely a cliché; this excellent book clarifies the complex relationship between emerging technologies and communities by providing new models to guide and develop our thinking. Wenger, White and Smith explore the subtle interplay of tools, practices and communities, and in so doing takes our level of understanding about this fascinating subject to a new level.

Mark Bennett, formerly the Principal Advisor - Communities of Practice at Rio Tinto, and in 2009 founder of LearningCollaboration.com

Though communities of practice and other collaborative workgroups find substantial and evolving support online, the growing scope and diversity of emerging social technologies presents a challenge - how do you keep up? This book offers a solution, a new role - technology steward, someone within the organization who can manage, monitor, and recommend platforms for collaboration. _Digital habitats: stewarding technology for communities_ defines the role and serves as a thoughtful, comprehensive guide for communities, and for those who, knowingly or not, fill the stewardship role. Beyond that, this is essential reading for anyone who needs or wants a better understanding of social and technical aspects of computer-mediated community.

Jon Lebkowsky, Principal, Social Web Strategies. http://socialwebstrategies.com - http://weblogsky.com/

Digital Habitats
stewarding technology for communities

Etienne Wenger, Nancy White and John D. Smith

Published by CPsquare

Digital Habitats: stewarding technology for communities
By Etienne Wenger, Nancy White and John D. Smith

Publisher: CPsquare
5908 SE 47th Avenue
Portland, OR 97206

Book website: http://technologyforcommunities.com

To report errors, please send a note to digitalhabitats@cpsquare.org

Book layout, illustrations and cover design by Michael Valentine
Cover painting by Randall David Tipton http://www.randalldavidtipton.com
Indexing by Sunday Oliver
Editing assistance from Jill Steinberg and from Peter and Trudy Johnson-Lenz http://johnson-lenz.com

ISBN 13: 978-0-9825036-0-7
1. Information technology - social aspects. 2. Computer networks - social aspects. 3. Internet - social aspects

Printed and bound in the United States of America

Print History: First Edition, August 2009

Contents

List of figures

Preface
The report update that became a book

The story of this book begins in 2000. The U.S. Federal Government's Council of CIOs commissioned Etienne Wenger to study Internet technologies designed to support communities of practice. His report was a broad survey of technology products in use at the time. It was made available in the spring of 2001. At the time, few products had been designed explicitly for communities of practice; a majority had been designed for other purposes such as instruction, collaboration, document storage, or conversation, and had been adopted by communities for their own use in learning together. The report compared the functionality of these different tools from the perspective of communities of practice and proposed some fundamental dimensions to make sense of the market.[1]

A changing landscape

Since 2001, technology-enabled communities of practice have proliferated, and the field of technology for communities has exploded. Digital habitats where communities dwell have changed remarkably. Many readers asked for an update of the original report. In early 2004 we began working on an update (joined by Kim Rowe for the first nine months). We started work on a revised report to reflect three areas we saw changing:

- **The evolution of the community support tools market.** Digital habitats are richer. Many of the products described in the first report had disappeared, merged, or morphed in interesting ways. New products and types of technology resources had emerged, with a growing number that were explicitly designed for communities of practice. Communities were adopting many more designed for other purposes. We were seeing an explosion of new tools, including a whole new way of defining tools known as Web 2.0. These tools facilitate the convergence of content and networks of people, creating new possibilities for communities to develop and grow.

---

1. Etienne Wenger, Supporting Communities of Practice: a survey of community-oriented technologies, http://www. ewenger.com/tech (accessed April 1, 2005).

- **Broadened mutual influence of community and technology.** Since the first report was produced, we found that interaction in digital habitats had advanced. Technology was being incorporated more deeply and broadly into the regular life of communities. The boundaries between tool selection, configuration, facilitation, and design were increasingly blurred. Even the questions people asked about technologies had evolved. Communities' conversations were becoming the vehicle for the evolution and even the development of technology. Community and technology were evolving in interwoven ways even more than before. The market—both proprietary and open-source—and the technologies in use were changing our view of community. They seemed to be transforming the very concept of community.

- **Deepened experience with technology.** We know more about digital habitats. Communities have accumulated vast experience using many different technologies. They have combined tools in new ways to serve their needs. As members, leaders, advisors, support staff, and students of communities of practice, we have observed successes and failures, and continue to appreciate both the opportunities and the limitations of technology.

We observe communities now facing larger and more complex technology choices for the creation of their digital habitats. Technology for community use has become an important area of practice, and one that needs to be developed and nurtured to yield its full potential. Awareness of these developments is shared by a growing number of people.

Changing us, changing the book

As we began, we did not know this report was going to become a book. Convinced that this project needed to be done but working without funding, we tried to make a virtue out of our slow pace by incorporating our own experiences in our reflections on the field as a whole. Likewise, we did not anticipate the extent of the fruitful learning about our own practices, passions, and interests generated by the dynamic, exciting, fun, and sometimes frustrating or problematic interplay between technology and community.

As a part of a developing community focused on community and its intersection with technology, we offer this book as a resource to help move the conversation along. Working together for the last three years, the three of us have become a little community of practice, bringing together different backgrounds, styles, and connections to our collaboration. Each of us comes from a different angle. Etienne brings an abiding focus on communities of practice and learning theory. Nancy adds the online interaction element with a longstanding interest in facilitation. John stirs it all up with experience both managing digital habitats and coaching community leaders. While the three of us think and work very differently, we're all focused on the same audacious goal: to contribute to the world's capacity to learn.

In addition to our shared writing project for this study we have collaborated on a series of other projects over the years—all using many of the tools and technologies we discuss here. Many of these projects have occurred in or around CPsquare[2], the community of practice on communities of practice, ranging from the community's launch, its online workshops, ongoing conversations (such as the annual "shadow the leader" cycle, one of which is referenced in Chapter 1), and many small conferences both face-to-face and online. We've used, reflected upon, and abandoned different tools, technologies, and practices. We've "looked over" each other's shoulders and learned from and with each other.

While we met for three wonderful face-to-face work sessions, most of our interactions have been at a distance. We've had lots of phone calls (about which we owe our families much gratitude for their patience and tolerance). Despite living in the same time zone

---

2. CPsquare, www.cpsquare.org

on the west coast of North America, our travel and commitments made it so we couldn't always work together at the same time. We worked online through Google Docs. We worked offline and exchanged documents via email with markups, comments, and annotations, becoming familiar with each other's file-naming conventions and quirks. When we had conference calls on a phone bridge, John and Nancy used a chat window in Skype to write notes during the conversation. (Etienne paced with a cordless phone!) Often, Etienne participated in his car on a cell phone while his family slept in a vacation cabin or his son was at gymnastics. John emailed the notes to all three of us so we could track our tasks. We tagged a lot of resources using http://delicious.com/tag/technologyforcommunity. Both Nancy and John blogged, often asking their readers questions about this work.

Our work and learning intertwined into ever-denser strands as time passed, influencing our own practices. As we discussed and chewed on ideas, we lived them with our communities and clients. We gained a new vocabulary for discussing the increasing influence of technologies on communities and communities on technologies. We connected our learning to our ongoing work.

Jump in!

From modest roots comes this offering from our minds, our experiences, and our hearts. It has sharpened our awareness of and belief in the importance of the interplay between technologies and communities. This interplay is clearly part of the learning challenges that our world faces today. In this spirit, we offer this work as a contribution to a broader conversation of which we are a part. We invite you to join us in this journey of discovery as we learn more about how to work, learn, and play together in our highly technology-mediated world.

Waving from across space and time!

Etienne Wenger, http://ewenger.com, North San Juan, California
Nancy White, http://fullcirc.com, Seattle, Washington
John D. Smith, http://learningalliances.net, Portland, Oregon
USA
June 2009

Acknowledgments

Our collaboration and conversations with a broader circle of fellow travelers also set the context for our work. We are involved in numerous communities and networks where people care about the topic of this book and we are deeply indebted to both the communities and the individuals who have allowed us to share their stories, torture them with early versions of the book, and generally learn together. These social connections and contexts have shaped the book. The cases we have looked at were largely determined by the network of people we had access to. At the same time, this network allowed us to share portions of the book as we were writing it, gaining invaluable insights from the comments we were receiving. This book therefore reflects our membership in our various communities, especially in CPsquare. We send special thanks out to Alice MacGillivray, Allison Hewlitt, Amy Lenzo, Andrew Mahon, Barbara Simonetti, Beth Kanter, Betsy Lowry, Beverly Trayner, Bill Bruck, Bob Doyle, Bronwyn Stuckey, Caren N. Levine, Christine Rizzuto, Dave Makowski, Dave Peloff, Denham Grey, Ed Mitchell, Elisabeth Davenport, George Siemens, Jack Merklein, Jay Cross, Jenny Ambrozek, Jill Steinberg, John Barben, Joitske Hulsebosch, Joseph Cothrel, Julie Schlack, Kim Rowe, Lee LeFever, Lilia Efimova, Lori R. Givan, Lucie Lamoureux, Marc Coenders, Mark Bennett, Mark Hammersley, Matthew Simpson, Michael Harrison, Michael Valentine, Miguel Cornejo, Noel Dickover, Patti Anklam, Peter+Trudy Johnson-Lenz, Robert Scoble, Robert Tollen, Shawn Callahan, Shirley Williams, Soren Kaplan, Stephen Gance, Sue Wolff, Sus Nyrop, and Tony Burgess.

Finally, without the support of our families, we would never have been able to complete the project. Deep thanks and love to them.

Itinerary
A reader's guide

This itinerary provides a brief overview of the book's structure. It has two purposes: to clarify our intentions in the way we wrote the book and to enable you to focus and skip around according to your needs.

How we envision you, the readers

In the preface, we've shared a bit about ourselves. But as we wrote the book, we have had to guess about you, based on what we know from our friends and colleagues who care about communities, technologies, and digital habitats. How you use what we offer will depend on who you are. We see three overlapping groups as being most interested in this book. Which best describes you today? Which one might you be tomorrow?

"Deep Divers"

We envision you as being interested in exploring the connections between technology and community more deeply through the application of learning theories to practical situations such as the use of technology by communities of practice. You are keen to create connections across disciplinary or practice boundaries.

"Attentive Practitioners"

You are interested in developing your practice, whether technology plays a major or minor role in your work. You are interested in learning with your fellow practitioners. You are searching for better ways to serve your communities and need to be able to talk about technology stewarding, demonstrate its value, and even communicate effectively about your identity in the role. You may describe yourself as a community leader, facilitator, sponsor, or coach, or as a technologist who supports, builds, or designs tool for communities. While you have a bias for action, you have a stake in both practice and theory.

"Just Do It-ers"

You are action-oriented, making things happen. By choice or assignment, you are tasked with supporting your community's technology. You get right down to figuring out how you are going to do the job. We have tried to include enough tips and practical chapters that you will be able to skip around and find material useful to you without necessarily encumbering your action orientation with the more conceptual aspects.

We have attempted to weave your three perspectives throughout the book on the assumption that there is value in combining them. We make specific recommendations and provide practical tips where we can. We also develop more general frameworks and ways of approaching problems that you can apply to your situation. Because so much of the language around technology is changing or is new, we have included a glossary at the end of the book.

What this book is not

Whatever your perspective is, a caveat is in order. While we hope that this book will be useful to those who need to select technology for their communities, it is not a guide to selecting technology. In this regard, we should state up front two things that this book is not:

- **It is not a shopper's guide to technology products.** The 2001 technology report that got us going on this book described technology products in detail, but this book does not attempt to do that. With the rate of change in the market, product reviews and recommendations today are stale and irrelevant tomorrow. Details about software and services mentioned in a book can change from one day to the next. We don't offer directives and advice about what software to get or reviews of specific products. When we do mention specific products it is to illustrate a point, not as a review or recommendation of the product itself. Instead, we offer a foundation that can live and evolve with both the needs of communities and the technology market.

- **It is not a roadmap to technology selection.** The book is not structured as a guidebook—with the first chapter describing the first step, the second chapter the next step, and so on. The order of the chapters does not reflect any process or sequence of steps for an activity. (In fact, our early focus on the technology landscape would be in total contradiction with our recommendation to focus on community needs first.) Rather, this book is about the identification, description, and practice of an emerging function. The order of the chapters reflects the need to build a useful repertoire of concepts, models, tools, and practical tips for those who undertake the role of stewarding technology for communities and its complex set of activities.

How we structured the book

While intending to write something useful for all three types of readers, we also realized that you have different needs. Some of you may read the book straight through; others will hop around. So we describe here how we have clustered the eleven chapters into four main parts.

Part I – Introduction (Chapters 1-3)

The first three chapters frame the book and define the notion of technology stewardship intellectually, historically, and practically.

- **_Chapter 1: How do we understand the notion of community?_**
  We clarify our perspective on communities by analyzing an example of an online discussion group as a "community of practice."

- **_Chapter 2: How are communities and technology related?_**
  We retrace the recent history of mutual influence between communities and technologies.

- **_Chapter 3: What is technology stewardship?_**
  We introduce the notion of _technology stewarding_ as an emergent form of leadership in communities.

Part II – Literacy (Chapters 4-6)

These chapters offer three models for thinking about technology in communities. These models are meant to help tech stewards "read" situations and propose courses of action. They constitute a kind of "literacy" of the function.

- **_Chapter 4: What aspects of technology should a steward consider?_**
  We distinguish four perspectives on technology: tools, features, platforms, and configurations. A "configuration" refers to the entire set of tools and features a community uses, whether these are bundled into one or many "platforms." The focus on a community's full configuration should be the perspective of a technology steward. It also places issues of tool integration at the core of a community perspective on technology.

- **_Chapter 5: Why are various tools useful for communities?_**
  We try to make sense of the tools that we see in the digital habitats of communities. We propose to understand how these tools help communities in terms of addressing three polarities that are fundamental dilemmas for communities: togetherness and separation across time and space; interacting and publishing; and individuals and groups.

- **_Chapter 6: What are patterns of community activities that technology can support?_**
  We identify nine different orientations to make sense of how a community might rely on technology. An orientation is a pattern of activities that has distinct implications for technology—a typical mix of activities and concerns through which members experience being a community and learning together. Describing each orientation and its technology implications makes for a long chapter, so you may want to pick and choose.

Part III – Practice (Chapters 7-10)

These chapters focus on the evolving practice of stewarding technology.

- ***Chapter 7: What are the contextual factors that a steward needs to take into account?***

  We explore a range of factors beyond the activities of the community that will affect the decisions of technology stewards.

- ***Chapter 8: What are the various strategies for providing a community with technology?***

  We identify seven different strategies for putting together a technology configuration for a community. These strategies reflect various levels of financial and technical resources available to a community.

- ***Chapter 9: What does it take to continue stewarding technology in use?***

  We describe the work of technology stewards as they support the use of technology in the day-to-day life of their communities. As communities evolve, new needs arise, tools are used in unexpected ways, members bring in their experience, and new technologies make their way into the practice. Stewarding technology in use is the practice and art of paying attention to this process and finding ways to support it—whether this work is visible or takes place in the background.

- ***Chapter 10: What do I do now?***

  This is an "action notebook," which summarizes the practical implications of each chapter in the book. As a "how-to" guide, it is organized as a series of steps. With a series of checklists, tables, and questions, it outlines what to keep in mind when you go about stewarding technology.

Stewarding is not a sequence of tasks that can be reduced to a recipe. Because a digital habitat is part of the life of a community, choosing technology, installing it, and supporting its use requires understanding and improvisation.

Part IV – Future (Chapters 11-12)

The final chapters are two parts of an essay on the future of technology stewardship.

- ***Chapter 11: Where is the interplay between community and technology going?***

  We return to the theme of Chapter 2—the mutual influence between communities and technologies—and projects this theme into the future by looking at current trends at the intersection of community and technology.

- ***Chapter 12: How should technology stewards develop their practice?***

  We use these trends to frame a "learning agenda" for the practice of technology stewarding that outlines a learning agenda for technology stewardship as an emerging practice with implications for communities at various levels of scale—from very local technology decisions to the enabling of communities that can span the globe.

This look at the future is not about predictions or action items. Rather, it is an invitation to imagine together how we can advance the practice of stewarding technology for communities, in the service of our communities and the world.

Where you should focus

We hope you will recognize yourself in the introductory description of the role of technology steward in Chapter 3.

- "Deep Divers" may be particularly interested in the glimpses of theory and history offered in Chapters 1 and 2 and in the reflections on the future in Chapters 11 and 12.

- "Attentive Practitioners" will appreciate the models in Chapters 4 to 6 as resources for sharpening their thinking and anchoring their practice.

- "Just-Do-It-ers" may want to start with Chapter 10, and only go to other chapters when they need additional information (the relevant chapters are indicated for each step). In terms of other chapters, "Just-Do-It-ers" will prefer the more concrete material in Part III (Chapters 7 to 9) as well as the concluding remarks in each chapter in Part II, which focus on implications for practice.

The online component of the book

To address the very real need to share ideas about specific and changing technologies, we have created two online spaces. One is a group blog at http://technologyforcommunities. com, where we share some of the worksheets and news related to the book. The other is a tools wiki (http://technologyforcommunities.com/tools) where we have pages for specific types and combinations of tools. We invite you to build on that wiki and help the larger community of people interested in technology for community to learn together, day to day. The Action Notebook (Chapter 10) is online at http://technologyforcommunities. com/actionnotebook.

Part I:
Introduction

Communities of practice: a glimpse of theory

In this chapter we talk about the perspective that this book uses to explore digital habitats — that of learning together in communities of practice. We illustrate this perspective by projecting it on a simple email list. Then we draw some implications for understanding technology from this perspective.

Before we delve into the topic of the book, we thought it might be useful to articulate the perspective we use to look at the interactions between a community and the technology it uses. For us, what is most interesting about the interplay of community and technology is our ability to learn together. In particular, when we talk about how technology enables community, we are particularly focused on *communities of practice*, communities where the learning component is central.[1]

---

1. A more complete formulation of the learning theory can be found in: Etienne Wenger, Communities of Practice; Learning, Meaning and Identity (New York: Cambridge University Press, 1998). Its implications for supporting learning in organizations are explored in: Etienne Wenger, Richard McDermott, and William Snyder, Cultivating Communities of Practice: a Guide to Managing Knowledge (Boston: Harvard Business School Press, 2001).

By learning, we do not mean just book learning, or classroom learning, or even e-learning. We see learning as an integral part of life. Sometimes it demands an effort; sometimes it is not even our goal. But it always involves who we are, what we do, who we seek to connect with, and what we aspire to become.

For our purpose, learning together forms a valuable perspective on the communal aspects of technology. It is more demanding of technology than keeping a list of friends or exchanging messages: it implies that technology will help us find learning partners and engage with them meaningfully. How email can contribute to the formation of a community of practice is a more specific, and therefore more demanding, question than whether email allows people to communicate. Blogging, for example, is often described as a public journaling technology. Yet it also gives people new ways to discover what they have in common, possibly leading to the formation of new communities.

We are not claiming that learning is the driver of all communities, certainly not as a concerted intention. Still, we believe that a large part of what makes interactions on the Internet attractive and productive is the ability to experience "learning friendship," as our colleague Marc Coenders calls the process of participation in communities of practice.

Seeing community in technology

We will illustrate the community of practice perspective with an example. We will apply it to a group of patients who mainly uses a very simple technology, an email list, to communicate. The story of this group is briefly outlined in the story box entitled "MPD-SUPPORT-L: *email list or community of practice?*" (For simplicity here, we will just call this group the MPD community.)

The question we propose to address is: What do you see when you look at this email list as the main platform for a community of practice?

To describe the way in which the use of an email list has opened a space for "learning together," we focus on three fundamental dimensions of a community of practice: domain, practice, and community.

The "domain" dimension

In coming together in an online conversation, MPD community members express something fundamental they have in common: the challenge of dealing with a potentially lethal but manageable blood disorder. Attention to something that members really care about is an essential aspect of a community of practice. For a community to form, the topic must be of more than just a passing interest.

The first observation we would make is that the MPD community has opened a space for exploring a specific *domain* of inquiry – in this case, the patient perspective on a family of related diseases. Sustaining a process of learning together over time depends on the definition of such a shared domain. It provides an identity for the community – a set of issues, challenges, and passions through which members recognize each other as learning partners.

The MPD community has established the patient perspective on these disorders as a legitimate focus for collective learning.

The domain inside: The definition of the domain is not without potential controversies. When an MPD member suggested giving up on western medicine and turning to natural remedies like beet juice, a heated debate ensued. What if someone followed this advice and died? What is the proper focus for the community? Should those interested in natural remedies open their own list? In this case, the community did not split, but the definition of its domain – and the question of its identity – was shown to be a potentially contested terrain.

The domain outside: A community's work on its domain often has significance beyond the immediate members. The MPD community places the patient

---

MPD-SUPPORT-L: email list or community of practice?

Myeloproliferative disorders (MPD) are a family of relatively rare blood diseases that are not curable but not lethal either if managed with proper medical care and lifestyle changes. After being diagnosed with a form of MPD, Robert Tollen started an email conversation with a physician friend who had also been recently diagnosed. Over time, other acquaintances asked to join the conversations. In 1996, Robert moved the conversation to an AOL listserv so that more people could benefit from the exchanges. Today, the listserv has more than 2,500 subscribers worldwide. Living with MPD is still very challenging, even if diagnosis has become much more precise and new treatments have been discovered. The group continues to share information about various aspects of MPD and provide real-time support for patients and their families. Their primary orientation to learning together is still an *asynchronous*, open-ended conversation, where participation involves reading and sending email. They still use the AOL listserv with a simple web page for open subscriptions, archives of all the exchanges since 1996, and the results of a comprehensive survey that depicts what living with the disease is about. Robert himself readily tries to adopt new alert and search tools to gather information about treatments and medical research on behalf of his community. In the last few years he has also introduced new tools. The community has a Frappr map site for member pictures and locations. A QuickTopic site holds hundreds of questions and answers and a shared file folder. Still, Robert is careful to avoid imposing new tools on the community as a whole. Not all MPD patients are technology savvy. All that members really need to participate is an email address, and he wants to keep it that way.[2]

Main site: http://www.mpdsupport.org

Q&A repository: http://www.quicktopic.com/32/D/HpPngyBqZ9LqH.html

Map: http://www.frappr.com/mpdsupport

---

2. We became aware of the MPD-Support community because the father-in-law of a friend benefited so much from its remarkable level of emotional and practical support, in addition to the scientific information that it provides. As a result of that initial contact, we invited Robert Tollen to visit with members of our CPsquare community once a month for a year to talk about his community and his experience as convener and coordinator (a regular learning activity we call "shadow the leader"). As of January 2008, the community email list had 2651 subscribers from 42 countries, a majority from the US, then the UK, Australia, and Europe.

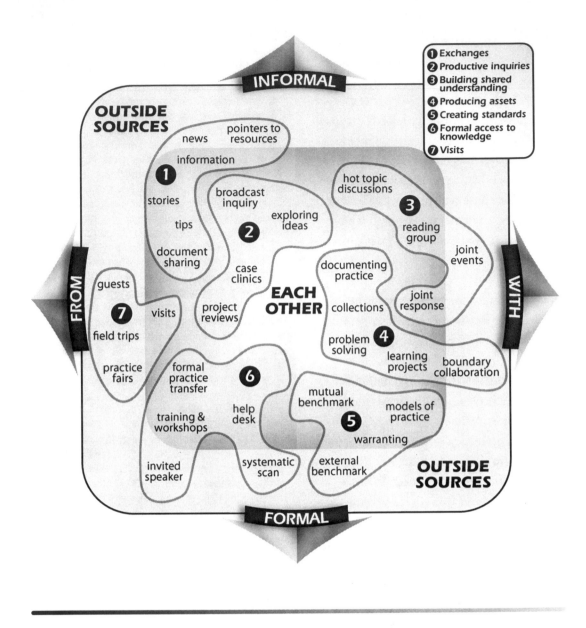

Figure 1.1. The range of activities in which communities of practice engage.

perspective on the disease in the public sphere. Some doctors and researchers have become aware of the community and have subscribed to the email list to witness the patient perspective. The community has the potential of creating a new relationship among patients, healthcare practitioners, and researchers. In fact, because of his leadership in the community, Robert was invited to give the closing keynote at a research conference.

The "practice" dimension

Because myeloproliferative disorders are lifelong afflictions, patients have to develop a practice of living with their disease. They can truly be called *practitioners*, not in the professional sense, but in the sense of sharing a practice. Their practice includes all the activities and techniques for coping with a life-threatening, chronic disease: going to the doctor, interpreting tests, understanding treatment options and prescriptions, accepting the diagnosis, staying on a diet, taking medication, understanding symptoms, and, more generally as patients, dealing with life as well as facing mortality. Learning a practice is learning how to be a certain kind of person with all the experiential complexity this implies: how to "live" knowledge, not just acquiring it in the abstract.

The common and personal experience of dealing with a disease provides an incentive to interact, hear each other's stories, compare notes about symptoms, home remedies, and the healthcare system, and notice each other's silence or absence. Because of this common experience, even a simple communication technology like an email list can provide enough access to each other's practice to make the learning extremely relevant to the lives of the community members. Over time, with the combination of a living conversation, an archive, and a patient-oriented FAQ, the community has developed into a set of valuable resources for practitioners of the disease.

On one level, it appears members of the MPD community mostly exchange emails with each other. From the perspective of developing a shared practice, however, they engage in a fairly complex set of learning activities. Figure 1.1, from an upcoming book on learning in communities of practice,[3] shows the range of activities that communities of practice have been known to engage in. While the details of the full set are more relevant to community facilitation than to technology (and are therefore beyond the scope of this book), it is useful for those interested in technology to appreciate the richness of what is possible. It is instructive to use the dimensions of the figure to see that even in a simple email list one can recognize a substantial subset of the possible activities.

Learning *from* and *with* each other. MPD community members learn *from* each other's day-to-day experience. They share stories, personal experiences with treatments, and tips ranging from dealing with itching to dealing with hematologists. They encourage newcomers to get evaluated by a specialist at a university hospital at least once. They also learn *with* each other. They help each other understand the possible impact of a recently introduced treatment or a newly discovered genetic marker. They struggle to make sense of a diagnosis and prognosis. They discuss the value of various approaches to treatment.

Learning through *formal* as well as *informal* activities. In the MPD community, learning activities are mostly informal: spontaneous exchanges of stories and tips, questions and answers, discussions of hot topics. But there are also some formal activities. The community collects the most important exchanges as topical summaries on its website and

---

3. Etienne Wenger, A Social Discipline of Learning (forthcoming).

a list of patient questions on its QuickTopic page. Robert conducts a systematic scan of the web for relevant information. He maintains a series of automated searches for topics such as "Interferon" treatments or the "Jak2" chromosome marker to keep abreast of new research findings. The most formal learning activity that the community has undertaken was a large survey of patients in 2001, covering everything from medications to food allergies to lifestyle characteristics—even including the acceptance of mortality. It required specialized software for data gathering, interpretation, and reporting.

Learning from sources outside as well as inside the community. In addition to peer-to-peer exchanges, an important function of the MPD community is to provide a window onto the wealth of available information from outside sources such as scientific journals, websites, and relevant news stories. Robert's scan provides access to a lot of outside resources, but other members also regularly contribute links, news, and opinions from outside the community. These patients have access to new information at the same time as healthcare professionals. Physicians and researchers who are subscribed to the list sometimes communicate with Robert in private to help him ensure that the community reflects outside perspectives.

The "community" dimension

The MPD community brings together patients who are seeking learning companions – people from all over the world who can say, "We know what it's like to have a myeloproliferative blood disorder." The email list creates moments of togetherness and ways to "hang out." Socializing and learning are not necessarily distinct. In general, the atmosphere on the list is supportive and welcoming. The full name of the community, *MPD-SUPPORT-L*, suggests that a strong element of mutual support exists and sustains the community. Indeed, people will announce that they are going to the hospital for a procedure and discuss their fears. But the learning connection is just as salient as a process of community-building. The commitment to domain and practice acts as a key source of trust among members. How could 2,500 people from all over the world, who communicate only by email, be called a community? In fact, when they read each other's postings they can recognize each other as practitioners because they read with a practitioner's eye. Therefore, they participate with the expectation that what is shared is going to speak to their own experience of practice – that they will learn something meaningful.

Learning together depends on the quality of relationships of trust and mutual engagement that members develop with each other, a productive management of community boundaries, and the ability of some to take leadership and to play various roles in moving the inquiry forward.

Diversity and engagement. MPD community members live all over the world, are of different ages, and come from different walks of life. They have several variants of myeloproliferative disorder; healthcare systems and drug availability vary all over the world; and members may have quite different views about the disease itself. Although most members are patients, we have mentioned that some professionals and relatives also subscribe. Community based on practice need not imply homogeneity: diversity in community is a good learning resource. One member, who was an Australian priest, secured a job in the Vatican because he had learned that treatment of myeloproliferative diseases in Italy was more advanced than it was in Australia. As mentioned earlier, disagreements do crop up; in fact, they are an essential ingredient of collective learning. But "flaming behavior," such as harsh words between "hard science" and "natural" advocates, are relatively rare. Through a focus on practice, a healthy community acts as a social container to make disagreements productive.

Legitimate peripheral participation. With more than two thousand subscribers to the list, it is obvious that many of them are not posting regularly – the community would implode if they did. A few are very active; some post occasionally; and a great many only read. In web lingo, these readers are called *lurkers*. From a community of practice perspective, *lurking* is interpreted as "legitimate peripheral participation," a crucial process by which communities offer learning opportunities to those on the periphery.[4] Rather than a simple distinction between active and passive members, this perspective draws attention to the richness of the periphery and the learning enabled (or not) by it.

Often, people on the periphery are taking the time to familiarize themselves with the functioning and point of view of the community before jumping in. People who remain peripheral may also carry the community's learning to other communities. For instance, researchers want to understand the experience of patients so they can take this perspective into their own practice, but they understand that their status as researchers may disrupt the peer-to-peer dynamics of the community if they were to become active. And patients who do not post are not necessarily less engaged or affected. In many cases, the intensity of the learning of these peripheral participants only comes to Robert's attention when a relative writes to cancel the subscription of a deceased member. The automated goodbye letter can prompt responses such as:

> "I was subscribed for 7 years and we read the list everyday with anticipation, but I never posted. My mother and I were helped to an incredible degree by the information we received from your list. Thank you."

Such feedback produces some of the most moving and energizing moments for Robert: years of practical and inspirational connection to the MPD community provided a lifeline to someone out on the community's periphery – someone who was mostly silent but nevertheless deeply appreciative.

---

4. Jean Lave and Etienne Wenger, Situated Learning: Legitimate Peripheral Participation (New York: Cambridge University Press, 1991).

Leadership is an essential ingredient in a community of practice, whether formal or informal, concentrated in a few people or broadly distributed. There is no question that Robert's leadership is key to the success of the community: not only his management of the list, but also his legitimacy as a patient, his focus on practical learning, his constant scan of the field, and his extensive personal experience and knowledge. (Because he is so obviously knowledgeable, in many of his communications he has to be careful to remind people that he is not a doctor and that his opinion cannot be taken as medical advice.) It is perhaps one of the risk factors in this community that it seems rather dependent on the leadership of one person, though a core group of active members shares some of the caring for the community.

We do not claim that all list subscribers would subscribe to our description of their community. When we started to talk with Robert, he thought of himself as just the owner of the MPD-SUPPORT-L list, not that what he was doing was so extraordinary. As our conversations proceeded, however, he came to agree that there was much more to his list than a list (and to his contribution than just "list ownership"). Today he would describe his group as a community of practice and his function as community convening.

Seeing technology through community

An email list works well for the MPD community's informal conversations. We can see many community aspects in it without the complications of more elaborate tools. But a large and rapidly growing array of technologies is available today, with varying potential to enable community. We can now articulate better why a community of practice perspective is a demanding, and therefore productive, way of looking at the interplay between community and technology.

Each one of the three dimensions – domain, practice, and community – places demands on technology; conversely, technology opens up new facets of each dimension:

Domain. How does technology enable communities and their members to explore, define, and express a common identity? To see the landscape of issues to address, and then negotiate a learning agenda worth pursuing? And to project "what they stand for" and what it means to them and others? Does technology allow communities to figure out and reveal how their domain relates to other domains, individuals, groups, organizations, or endeavors?

Practice. How does technology enable sustained mutual engagement around a practice? Can it provide new windows into each other's practice? What learning activities would this make possible? Can technology accelerate the cycle through which members explore, test, and refine good practice? Over time, can technology help a community create a shared context for people to have ongoing exchanges, articulate perspectives, accumulate knowledge, and provide access to stories, tools, solutions, and concepts?

Community. How can technology support an experience of togetherness that makes a community a social container for learning together? Can it help people find each other and reduce the sense of isolation? Does it reveal interesting connections and enable members to get to know each other in relevant ways? Can it enhance the simultaneous interplay of diversity and common ground? Does it allow various people and groups to take initiative, assume leadership, develop roles, and create subgroups, projects, and conversations?

Technology extends and reframes how communities organize and express boundaries and relationships, which changes the dynamics of participation, peripherality, and legitimacy. It enables very large groups to share information and ideas at the same time as it helps smaller groups with narrower, more specialized and differentiated domains to form and function effectively. It allows communities to emerge in public, opening their boundaries limitlessly, but it also makes it easy to set up private spaces that are open only to members. It affords many ways to limit access, expressing intimacy or privilege, or it can greatly enlarge a group's periphery. A person who comes across a community site as a result of a search engine, combs it for ideas and information, and never comes back is part of the largest periphery.

Communities of practice offer a useful perspective on technology because they are not defined by place or by personal characteristics, but by people's potential to learn together. Unlike the trajectory of a team that's planned from the start, communities unfold over time without a predefined ending point. Communities often start tentatively, with only an initial sense of why they should come together and with modest technology resources. Then they continuously reinvent themselves. Their understanding of their domain expands. New members join, others leave. Their practice evolves. The community's technologies need to support this intertwined evolution of domain, community, and practice – a very challenging goal.

We use the communities of practice perspective to understand digital habitats – where community and technology intersect. It helps us focus on how communities use technology, how they are influenced by it, how technology presents new learning opportunities for communities, and how communities continue to assess the value of different tools and technologies over time, and even how communities influence the use of technologies. The close, voluntary collaboration in communities enables their members to invent and share new uses for the technologies at their disposal. Communities often play a key role in the dissemination and appropriation of new technologies. The social lens needed to understand the way communities of practice use technology can be helpful in understanding many of the issues faced in other groups where the learning component is less salient – social networks, virtual teams, friendship groups, conversations. Therefore, we believe the discussions in this book will be useful to people who are supporting technologies in all sorts of groups and networks regardless of whether or not they identify themselves as communities of practice.

Technology and community: a glimpse of history

*H*ere we begin our argument about the mutual influence of community and technology in digital habitats by looking back at some examples in recent history, first looking at how community was a factor in the development of new technologies and then at how technology was a factor in the development of new communities.

In 1973, David Woolley was a 17-year old University of Illinois student working on the design team for PLATO, a new computer-based learning platform. He noticed how difficult it was for people in the PLATO development community to track bugs. So he created PLATO Notes, a way to tag and track software bug reports more systematically. That same year, Doug Brown, another member, developed a chat facility for PLATO to provide a new way of having informal, peer-to-peer communication. David and Doug noticed their group's needs and introduced solutions that enabled greater collaboration in the PLATO development community: they were stewarding technology on behalf of a community.[1]

PLATO Notes and chat were online tools explicitly designed for one community, helping it improve its own practice (software development). David and Doug's intention to solve specific communication challenges drove technological development. By inventing new tools for working online, David and Doug expanded the ways of interacting that were available to their community. At that time *mini-computers* with hard-wired terminals and modems, such as those on which PLATO was built, were mainly used for "calculations" or for things like storing accounting data. As other tools like access lists and threaded discussions were invented, they opened additional collaboration options and were in turn adopted by the PLATO community. Needs spawned technologies and technologies inspired new practices, intertwining the two over time.

Others outside the PLATO community took note of what these designers had done and found the tools useful. This launched a series of developments that supported collaboration and community far beyond the boundaries of the PLATO community. PLATO is recognized as the inspiration for many subsequent collaboration platforms, including Lotus Notes and many other collaboration and bulletin board systems. Those systems not only built on PLATO's original design, but they also expanded the idea and practice of a working community. That idea would later flourish as early *community platforms* when the World Wide Web made the Internet widely accessible.

Technologies ➡ communities

The interplay of technology and community has a long history. Postal mail allowed scientists to exchange ideas and data for centuries. The telephone opened the doors to instant communication for those less inclined to communicate in writing. When the ARPANET began to connect academic and scientific communities, technology was allowing the rapid distribution and sharing of information on a much larger scale. In 1972, email software was introduced, giving non-programmers practical access to this new means of communicating and sharing information. With this possibility, people began to invent new ways to learn and work together. Platforms like PLATO could be developed outside the boundaries of hardwired computer networks. Early adopters started to form *virtual communities*. They overcame difficult and costly technology challenges, hooking up their modems at high per-minute rates because this new form of connection between individuals offered something important. It was a new challenge to figure out what it meant to be together online.

---

1. David Woolley, PLATO: The Emergence of Online Community, http://thinkofit.com/plato/dwplato.htm (accessed January 2007).

EIES

In 1977, the Electronic Information Exchange System (EIES)[2] became one of the first computer conferencing systems explicitly designed to support online group work. The design of EIES was built on a vision of community and technology as complementary, drawing on many technical and social innovations. The expertise brought to bear on the design by Murray Turoff and S. Roxanne Hiltz included not only computer science, but also communication systems, expert decision-making processes, group work, and ways of looking at communities and networks.[3] EIES supported some of the first online communities of practice, focusing on widely different domains: general systems theory, futures research, and social network analysis. Legislative researchers in 25 states shared their questions and each state's experience on topics from energy stamps to sludge disposal. Other online communities of practice shared their knowledge about devices for the disabled, set standards in the electronics industry, and dealt with rapid growth in small communities in the western regions of the United States.

Bulletin boards and email lists

Ward Christianson's invention of the first electronic bulletin board in 1977-1978 offered online collaboration to a whole new segment of people. With the explosion of email, email lists provided another group online interaction environment. With the development of email list technologies such as Listserv™, moderation functions were explicitly designed into the software to support newly-recognized functions: approving or rejecting memberships and messages, and enforcing group norms and agreements. The technologies started to provide functions for newly emerging social practices.[4]

Usenet

Soon after that, Usenet or "user's network" (also called the *newsgroups*) ushered in mass collaboration and conversations on the Internet. It was the first widespread peer-to-peer network, starting as a file sharing utility built on the TCP/IP protocol, and rapidly expanding to discussions, as people realized the power of connecting with each other, beyond their usual geographic reach.

From three sites and about two postings a day in 1979, Usenet grew rapidly. It proved so popular that by 1981, Mark Horton, a U. C. Berkeley graduate student, and Matt Glickman, a high school student, rewrote the access software to help deal with the volume. The volume and quality again became an issue in 1984, so the community began to develop software and processes to add moderation capabilities to the system. Community use patterns were creating the impetus to evolve both the system infrastructure and use practices.

---

2. Electronic Information Exchange System (EIES), http://www.njit.edu/v2/CCCC/eies/eiesinfo.html (accessed April, 2008).
3. S. R. Hiltz and Murray Turoff, The Network Nation, Revised Edition (MIT Press,1993).
4. H.E Hardy, The History of the Net (Master's thesis, School of Communications, Grand Valley State University, 1993), http://www.eff.org/Net_culture/net.history.txt (accessed May 5, 2007).

The "Great Renaming" in 1986-1987 was the Usenet community's response to its own success. The pressures of volume, the need to subdivide larger topic groups into smaller subareas, and the difference in group norms all triggered a massive and fundamental reorganization of Usenet. This was not only a reorganization of categories, but also a way to move groups that were exhibiting behavior unacceptable to other users off to their own "talk" area. It introduced the formal distinction between moderated and unmoderated groups. This pattern of experimentation, adoption, growth, and modification shows how many people were attending to the technology needs of the group, with the more technology-savvy members of the community taking an important lead in the design and implementation process.[5]

The Well

The Well (http://www.well.com) was started as an online community in 1985, not just resting on the technology to connect via telephone modems, but also with a clear intention for community with boundaries, norms, and agreements and a customized technology platform designed to support distinct topical subgroups. There was a system of hosts for each topical area. Members debated the expectations, rules, and agreements that should guide their behavior. Some members held regular face-to-face parties to celebrate their new-found togetherness, particularly in the California Bay Area where The Well infrastructure was housed. As the community grew, members contributed money to buy servers, and helped write and improve upon the community software. All the while they celebrated weddings, births, and deaths in the community, they fought with each other, they made an online home for the fans of the Grateful Dead band ("Deadheads"), and they became the archetype for the emerging concept of *online community*. They grew not only their community and their infrastructure but also the very form they epitomized.

Large-scale commercial bulletin board services

Seeing a niche for commercial products, pay-based dial-up bulletin board services such as AOL, Prodigy, and Compuserve[6] (with roots back to 1969) popularized the idea in the early 1990s that people could connect in all sorts of ways through a computer network: easy email, discussion forums, chat rooms, and buddy lists. Learning from The Well, they saw that personal computers had put the tools in businesses and homes and not just in universities. A revolution was underway. People in all walks of life could now exchange data, communicate, and collaborate across time and distance at a speed and volume never before possible.

The World Wide Web — Web 1.0

With the spread of the browser and the World Wide Web's hypertext interface in 1993, online access was further democratized, freeing networked communication from the

---

5. Usenet Newsgroup History, http://www.livinginternet.com/u/ui.htm (accessed February 14, 2007).
6. McAtee, AOL: A History, http://iml.jou.ufl.edu/projects/Fall2000/McAtee (accessed February 14, 2007).

proprietary systems that had popularized it and making access a simple matter of clicking and browsing. Although it was conceived of as a document publication system, The Web quickly became the Internet's dominant interface. It provided easy access, through its graphical and clickable user interface, quickly becoming a common medium for interaction. It enabled a new wave of innovation and invention for individuals and communities across the globe. A whole generation of companies attempted to satisfy these emergent possibilities with software, infrastructure and ideas of their own. As a result, online communities could congregate on websites designed explicitly for this purpose. Having a website both for sharing documents and for interaction became a rallying point for many communities.

Web 2.0

The notion of a "read/write web" or Web 2.0 was popularized by Tim O'Reilly in 2005.[7] He brought attention to a new generation of social software tools and characterized them collectively as social media. As we have observed, the story of technology stimulating community is an old one: PLATO was already social software. But the new Web 2.0 era has greatly increased the rate of change and the influence of technology on communities. New tools offer ways for people to rapidly generate, visualize, and connect networks. Within those networks, they can convene communities to learn and collaborate around the world, around the clock, expanding the ease and reach of peer-to-peer interactions. The social nature of the Internet has been greatly expanded by the many new ways to publish, interact, express individual identity, and form groups.

Countless groups of every imaginable flavor, size, and color use hundreds of different technologies and tools to interact in all sorts of ways. Geeks commune on Slashdot[8] using a custom-built platform that handles huge numbers of users, page views, and separate conversations. Bloggers, irrespective of the software they use to read, post, or publish, meet to go on walks together. People raise funds and call out for tips on software use through Twitter.[9] Educators have regular meetings in Second Life,[10] a three-dimensional virtual world. Mobile phone "thumb tribes" text each other to suddenly "smartmob" at a physical location. Technology is fundamentally expanding the possibilities of what it means to "be together." This rapid rate of change is also changing what we understand communities to be and is an important issue we'll revisit in Chapter 11.

7. Tim O'Reilly, What Is Web 2.0: Design Patterns and Business Models for the Next Generation of Software, http://www.oreillynet.com/pub/a/oreilly/tim/news/2005/09/30/what-is-web-20.html (accessed February 14, 2007).
8. Slashdot Community, www.slashdot.com
9. Twitter, www.twitter.com
10. Second Life, www.secondlife.com

Communities ➡ technologies

If technology has enabled communities to form and to interact in new ways, it's just as much the case that communities have played a critical role in the invention of new technologies. Many of today's technologies and practices are the result of the innovative thinking that first occurred in specific communities of practice. Other technologies have been invented because someone recognized a need in a community was not being addressed. The development of the first forum software in PLATO was an example of both: a small, innovative community and a need that could be recognized only in that new community setting. Sometimes the generality of a problem is only recognized afterwards: in *The Cathedral and the Bazaar*, Eric Raymond observes that, "The best hacks start out as personal solutions to the author's everyday problems, and spread because the problem turns out to be typical for a large class of users."[11]

The Internauts

In many cases, the inventors of a technology were part of a community that incubated their inventions.[12] The Internet itself has community roots. As Barry Leiner, Vint Cerf, and other "progenitors" of the Internet recount, its very evolution happened in large part because of their community of "Internauts," as they called themselves. They thought, worked, played, and innovated together, building on their diverse perspectives and shared passion. Today's Internet and the technological underpinnings of packet switching emerged from this intense community.[13]

Physicists at CERN

The World Wide Web was initially conceived to support the work of a community. Tim Berners-Lee was working at the Geneva-based European Center for Nuclear Research (CERN), whose scientists were located both in Geneva and at sites across Europe. He proposed his hypertext-based access system as a way to enable a community of high-energy physicists to communicate and share documents easily across the sites. The system was tried at CERN, but it quickly became clear to Tim and his colleagues that, in serving their specific community, they had developed a system that had the potential for much broader applicability. By 1993, just three years after the first demonstration at CERN, Tim's evangelism and the creation of the browser Mosaic had made the web a worldwide phenomenon.[14]

---

11. E.S. Raymond, The Cathedral and the Bazaar, http://www.catb.org/~esr/writings/cathedral-bazaar/cathedral-bazaar/index.html (accessed July 19, 2007).

12. Jonathan Grudin, "Groupware and Social Dynamics: Eight Challenges for Developers," Communications of the ACM, 37, no. 1 (1994): 92-105, http://research.microsoft.com/research/coet/Grudin/papers/CACM1994.pdf (accessed January 23, 2007).

13. B.M. Leiner, Vinton G. Cerf, David D. Clark et al., A Brief History of the Internet, http://www.isoc.org/internet/history/brief.html (accessed January, 2007).

14. R. Zakon, Hobbes' Internet Timeline, http://www.zakon.org/robert/internet/timeline (accessed January 24, 2008).

Software developers and "the wiki"

The wiki was invented by Ward Cunningham,[15] one of the leaders of a community gathered around the idea of using *pattern languages* in software development. Cunningham's intention was to create the simplest possible website, where anybody who wanted to could easily edit any page. He invited a dozen trusted members of the community to experiment with a prototype and create some sample content. The wiki was such a good idea for this community that it soon resulted in exponential growth of traffic, even with a rather idiosyncratic user interface. As it matured as a community knowledge repository, people in the community could be heard brushing aside a topic of discussion by saying, "The answer to that question is already on the wiki." Over time the community devised new uses for the wiki, including discussions of the behavior of wiki users or the uses of the wiki itself. Ward used those discussions to inform its further development – to extract data from an activity log about which page was read or edited most recently or most frequently, or about the most active "reader" or page editor, and about other technical and social issues. As wikis were adopted by other communities, the idea was developed further to reflect other needs, including devising interfaces that were attractive to non-programmers. From its roots in the needs of a rather esoteric community of programmers, today's wikis, and the collaborative capabilities they offer, are an emblem of Web 2.0 technology and are a social force to be reckoned with.[16]

Open-source communities

Open-source software development is a good example of how communities can use technology to collaborate as well as invent and transform the tools that are available to meet their needs. The community that has formed around the Apache HTTP server is remarkable not only because it has built the dominant software for an essential function that powers the web, but also because it has developed lasting partnerships between organizations and individuals.[17] Such communities not only use but also adapt, extend, and in some cases invent the technologies that enable them to come together. As they share the fruits of their labor with the larger world, these tools are shaped and reshaped by other communities in turn.

Technology and communities: a productive intertwining

In the past, many of us may have thought about these stories as mostly about technological invention. We now see them as stories about community and technology in their ongoing interaction. As communities appropriate technologies, they "make themselves at home" in new ways and in new places. They shape their digital habitats and the tech-

---

15. Wikipedia, The Free Encyclopedia, "Ward Cunningham," http://en.wikipedia.org/w/index.php?title=Ward_Cunningham&oldid=223408111 (accessed July 15, 2008).

16. Ward Cunningham, personal communication, 2005. See Ward's original wiki and further information at www.c2.com.

17. Bylaws of the Apache Software Foundation, http://www.apache.org/foundation/bylaws.html. (accessed July 15, 2008).

nologies they contain through novel use, asking more of the technology creators and suggesting new directions for development. As tools get easier to use, more and more members participate in the shaping process, taking the interaction between technology and community further.

Our stories describe this intertwining of technology and community where each shapes the other—in obvious and not so obvious ways. We see it as a creative vortex, moving forward, picking up new ideas and practices, and throwing off others like a tornado. Technology fuels this vortex, and so do communities— shaping the technology to meet community needs and extending a community's connections at each step of development.

Sometimes this vortex is moving so fast that we cannot see the patterns that form

> **Responding to community inventiveness**
>
> Twitter, a microblogging tool, was picked up by a technology-savvy crowd participating in the South by Southwest conference[18] in 2007 (many of whom were participating remotely). The quick adoption created a critical mass of users that both encouraged experimentation and amplified new ideas during the conference. They invented specific practices that allowed Twitter users to direct a post to an individual (while still chatting in public) by using the "@ username" convention.
>
> The Twitter software developers immediately picked up on the pattern and segmented the "@ posts" into a separate page, labeled "replies," so a user could see who had sent them "@ messages." This interaction between the inventiveness of the community and the developers built upon each other—all within a matter of days.

and repeat again and again. Today's community leaders have so many tools to choose from, all the while contending with the intricacies of social interaction with diverse people from so many backgrounds. This requires the invention of all sorts of new practices. Even largely co-located communities are changed by their use of technology to share documents, augment face-to-face interactions, stay in touch between meetings, or make community announcements. Wide adoption of community-oriented technology is due to the fact that it expands the available infrastructure for something fundamental to our humanity: social interaction.

Shared DNA

The patterns of interactivity and connectivity enabled by these recently introduced technologies are in remarkable alignment with the ways communities function as a context for learning. They exhibit several common patterns, such as a balance between independence and interdependence, an emphasis on horizontal relationships, and dynamic boundaries.

In retrospect, looking back at the development of the Internet, it makes sense that these technology developments would have such a profound effect on the behavior of, and, indeed, on our very notion of community. The stories about David Woolley, Doug Brown, Mark Horton, Matt Glickman, Tim Berners-Lee, and Ward Cunningham are just a

---

18. South By Southwest Interactive, http://2008.sxsw.com/interactive (accessed July 15, 2008).

glimpse into the early days of innovation and stewarding that put computers to work in the service of communities. Their work built on earlier contributions of pioneers such as Vanevar Bush, Doug Englebart, and many others. After the technologists, others came in to steward the technology from a more social perspective. Peter and Trudy Johnson-Lenz, Howard Rheingold, Lisa Kimball, and many others actively explored, invented, and wrote about ways that technology could support community interactions. Each person, community, and technology contributed to this vortex.

Technology has changed how we think about communities, and communities have changed our uses of technology. These evolving digital habitats give us the chance to reconsider what we know about communities and to rediscover fundamental ideas in new settings – to explore and, in the end, to know the place for the first time, once again.

Technology stewardship: an emerging practice

Emerging from the convergence of technology and community is a new role, which we call technology stewardship. We introduce the role and describe it in this chapter. This new role implies new functions, practices and identity. The role is important in helping communities construct and live in suitable digital habitats.

The closeness and enthusiasm in a community of practice makes it easy to spread news. When people find a great new tool, they tend to tell everybody about it, urging their colleagues to give it a try. When they start to find a tool limiting and frustrating, their gripes soon spread across the community until someone else comes up with a practice or a new tool to cope with the problem. The interplay between technology and community in this ongoing pattern of use, adaptation, and dissemination of evolving practice can happen by itself, untended. But we often notice that a person or a group of people are paying attention to this process and influencing it. As we began writing this book, we realized that we know these people, and we wanted to write about what they do. This chapter introduces the role of *technology stewards* (or more affectionately, *tech stewards*), and the activities, characteristics, circumstances, and motivations of those who undertake the role. Maybe you'll recognize yourself.

A working definition of technology stewarding

Technologies present new opportunities and challenges to communities. As more communities choose to use technologies to help them be together, a distinct function emerges to attend to this interplay between technology and the community: we call it *technology stewarding* to suggest how these individuals take responsibility for a community's technology resources for a time. Technology stewarding adopts a community's perspective to help a community choose, configure, and use technologies to best suit its needs. Tech stewards attend both to what happens spontaneously and what can happen purposefully, by plan and by cultivation of insights into what actually works.

> **Stepping into the role**
>
> Amy Lenzo didn't know it at the time, but she became the tech steward for the World Café community. While helping improve the visuals and navigation of the static website of this global community of practitioners who use World Cafés in their work, Amy realized that there was a visual and navigational discontinuity among the other online tools the community was using. The more she got involved, the more she noticed the need to address how the community worked across their various tools. Soon she found herself helping create both visual and technological devices as well as practices to bridge across the tools.
>
> http://www.theworldcafe.org

Technology stewarding is both a perspective and a practice. It can be considered a collection of activities carried out by the individual tech stewards and as a role within the community. The perspective is a natural outcome of taking care of a community that's using technology to learn together. Adopting the perspective means becoming sensitive to many different social and technical issues that we examine in this book, and developing a language to give the perspective voice and precision. A good example of stewarding is noticing that a community has grown so large that many people don't know each other, so the tech steward sets up a membership directory in response. A different sort of stewarding happens when a community that has used a simple email list for years decides it wants more functionality and asks a small group of technology-savvy members to research potential technology options, beginning the whole phase of moving to a new platform. Good tech stewards provide the level of technical expertise needed by a particular community. Their role may be invisible until the community's needs call for attention to technology. Some communities never grow beyond their initial needs, so ongoing stewarding is limited. Others develop complex configurations that need constant and deliberate attention.

Technology stewards are people with enough experience of the workings of a community to understand its technology needs, and enough experience with or interest in technology to take leadership in addressing those needs. Stewarding typically includes selecting and configuring technology, as well as supporting its use in the practice of the community.

This definition is meant to clearly distinguish between technology stewardship and traditional IT support. By emphasizing the experience with the workings of the community, we emphasize the insider perspective that shines a very specific light on the potential fit between community aspirations and technology. This insider perspective also emphasizes the practices that a community has to develop to leverage technology.

Technology stewardship is something anyone can do. It does not require absolute expertise with technology, but enough to play the role—for instance, to see the potential usefulness of a tool or represent the community's needs to other technologists. Many people contribute to technology stewardship even though they would never introduce themselves as "community technology stewards." In some communities, technology is the focus of one individual or a small group. In other cases, the work can be shared more widely or even be dispersed across an entire community, as often happens in technology-oriented communities where almost everyone shares in the responsibility.

Stewarding technology should be treated as a team sport for two reasons. First, it helps to have a group within a community share the work—or at least share in the understanding of the role. Second, it helps to connect with other stewards (from whatever community) who can provide a larger context, offer support, share ideas, tips, and innovation, and help in pressuring a tool developer to address community needs. Still, many tech stewards struggle alone.

Tech stewards most often are members of the community they serve. They just happen to pay attention to technology issues in the community's life. Sometimes, they just provide a service to the community, such as resetting passwords or archiving records. But in many cases, technology stewardship is a critical part of community leadership, facilitating a community's emergence or growth. It becomes a very creative practice that evolves along with the community and reflects the community's self-design—the process by which a community "designs" itself as a vehicle for learning, which includes use of technology. Actual community membership is not a condition for the role, but it does confer it additional legitimacy.

Technology stewarding is often taken on in combination with other leadership roles in a community, in response to the needs of the situation. Indeed, community leaders may find themselves thrust into technology stewarding, whether they are ready or not. Whether it is a distinct role or a task carried out by a community coordinator, technology stewardship is part of community leadership. Tech stewards have to work with others to shape a community's direction and development.

For simplicity, in this book we refer to *tech stewards* talking about the role as if it were distinct and carried out by one person. The situation is usually more complicated.

Streams of activity

Technology stewardship involves several streams of activity. These streams can become more or less salient at various times, but they should not be thought of as a sequence. They mostly run in parallel and constantly inform each other.

Community understanding. The first and foremost activity of tech stewards is to understand their community and its evolution well enough to be able to respond to its expressed and unexpressed needs with respect to technology. This understanding of how the community functions includes its key activities, member characteristics, subgroups, boundaries, aspirations, potential, limitations, as well as its context. Achieving such understanding will require a combination of direct involvement, observations, and conversations with community members.

Technology awareness. With the community perspective in mind, tech stewards need to have enough awareness of technology developments to have a sense of what is available and possible. As we mentioned earlier, this awareness need not be as rigorously detailed or systematic as that of a technology expert, but it should enable a tech steward to recognize relevant opportunities and initiate further explorations. Technology awareness requires an informal but ongoing scanning of the technology landscape—through personal experience, playfulness, conversations, reading, or participation in technology-oriented communities.

Selection and installation. The combination of community understanding and technology awareness should enable tech stewards to help their communities make informed choices about technology. This involves both large and small decisions, such as selecting a whole new platform, choosing to upgrade to a new version of a tool, or even advising the community to settle for what is "good enough" at the moment. Sometimes tech stewards propose technology choices, but sometimes members bring tools for the community to adopt. When it is time to make significant decisions, a community usually pays more attention to both technology and technology stewarding—often raising the level of expectation for the role. The technical aspects of selection and installation may necessitate additional expertise to help in the process.

Adoption and transition. Selection and installation are only one half of the equation. Tech stewards also need to shepherd their community through the process of adopting (or rejecting) the new technology. When a community changes technology in a big way, planning and facilitating the non-technological aspects of the transition process is a substantial task. Tech stewards can play a critical role in taking their communities through the learning curve usually associated with technology adoption and transition.

Everyday use. Tech stewards need to integrate the use of technology into the everyday practice of the community as it evolves. This stream of activities, which is often less visible, involves all sorts of tasks, from the mundane to the sophisticated. It has technical aspects such as tool management, upgrades, access and security, and back-ups. It also has community aspects such as onboarding newcomers, discovering and spreading new practices associated with the use of tools, helping craft agreements about technology use, and building capacity for stewarding in others. These tasks require observing, listening, inventing, and teaching. They also help tech stewards maintain the ongoing understanding of the community necessary for seeing emerging needs and participating actively in its evolving self-design from the technology side.

Again, in all these activity streams, what distinguishes a community tech steward from other technologists is

Figure 3.1. The centrality of the community perspective in all tasks

placing the community perspective at the core of technology-related challenges such as scanning the landscape, choosing technology, and supporting its use (see Figure 3.1).

The role of technology steward

Communities looking for tech stewards may seek individuals who like to "play" with tools who are also interested in how the community's activities are carried out. Some communities may encourage technically adept individuals to join, recruiting them to the role.

They may find themselves calling on the services of tech stewards from other communities. In some cases tech stewards don't adequately attend to the community's technologies and may have to be replaced, particularly when one individual starts to use the role to control, rather than serve, the community.

While the role that tech stewards play depends on the communities they serve, there are some characteristics we have observed in practice:

About technology and practice. While knowledge of technology is a key asset, tech stewards pay attention to how technology is used to achieve community ends. The alignment of tech stewards with the values and direction of a community makes them able to contribute in ways that technical experts might not.

A broker. Tech stewards often serve as brokers between the community and the technical resources in its vicinity, such as an IT department, an open-source community, or a vendor's support organization. A broker is in a position to appreciate the concerns and knowledge resources of people who can't always talk to each other directly. Good stewarding involves knowing who might know what and bringing into their community relevant perspectives, ideas, or possibilities from other practices.

Part-time/voluntary/paid. The tech steward role is usually part-time—whether it is an ad hoc response to a need or a longer-term commitment. Since participation in most communities of practice is itself part-time and voluntary, the amount of time available for the role of tech stewards is also limited, even though some communities can seem like children with infinite appetites. In communities supported by organizations, tech stewards can benefit from release time, extra resources, coaching, or other support, especially while they are new in the role. Having the technologists in an IT department recognize the legitimacy of the tech steward's role will save wear and tear on voluntary tech stewards.

Occasional visibility. In periods of stability the role may not be very visible—until problems arise or new technologies are introduced. There is work to be done in the background, nevertheless. When technology problems arise or new technologies are introduced, the visibility of and demands on tech stewards can change dramatically.

Why take on the role of technology steward?

People take on the role of tech steward for very different reasons, from personal interest to curiosity to generosity. Others are thrust into the role without much of a choice, perhaps simply because of previous experience with technology. But why would anyone accept the role, much less volunteer for it? Although the role is important to the community as

Chapter 3. Technology stewardship: an emerging practice

described, tech stewards themselves need to be clear about the benefits of signing up for a role with a rather steep learning curve.

- **No one else is doing it.** Without a community member taking on some of the leadership function, the community's technologies may be poorly developed or shaped by others who can't do the job as well.

- **Satisfaction in serving.** The satisfaction of making a contribution to a community is a prominent motivation. It is the value of the community and its contribution to the world that ultimately is the biggest motivator.

- **Leadership opportunity.** Technology stewarding provides a very concrete area for leadership that shapes how the community "gets together" and how productive it is.

- **To learn and grow.** The role presents a way for individuals to stretch their technical skills and learning. The "learning atmosphere" of a community of practice can be especially helpful in this regard.

- **Reputation building.** Credibility as a tech steward can be a useful entrée into other leadership activities, providing access to community leaders and increased visibility.

> **From curious explorer to thought leader**
>
> Joitske Hulsebosch had a long-term interest in both communities of practice and technology. She started out blogging about communities, and then added layers about technology. Soon she found herself part of a group of NGO staffers looking to better understand the use of new social media for their communities and organizations. Joitske became a facilitator of this community of practice with a few others and organized face-to-face workshops and online experiments with new tools. She captured her learning in her blog and supported the group to blog their experiences too.
>
> Without a conscious plan, Joitske found herself recognized as a tech steward or guru in the NGO community. People looked to her to help them in their technology stewarding and sought her insights on new tools. Her own stewarding increased both her knowledge and her reputation, and she has now become a consultant in the field of communities of practice and social media.
>
> http://joitskehulsebosch.blogspot.com

Stewarding technology in diverse circumstances

The nature and relevance of technology stewarding depends on the community and its circumstances. The role presents different kinds of challenges, depending on the community's size, stage of development, diversity, level of support, membership age, organizational setting, and even interest in technology. A community that uses a Yahoo! Group[1] and a phone conference call bridge has different stewarding needs than one that uses a wide range of tools for complex activities. A community with a dedicated and experienced technology steward may seek a more complex configuration than one that has little technical expertise. Some communities mostly interact online. When technology is the only way for members to meet and work together, technology stewardship is especially important.

---

1. Yahoo Groups, www.yahoogroups.com

A key factor affecting stewarding is where the community is situated. A community within an organization is likely to be dependent on a particular information technology (IT) infrastructure. If a community is spread across many organizations, it will have to bridge IT infrastructures. If it lives outside the boundaries and resources of any organization, it will have to fend for itself. These circumstances place different demands on tech stewards.

Stewarding within an organization

An organizational context shapes technology stewarding because it brings with it both resources and constraints. It can offer technology and support along with rules about tools and practices that can or cannot be used. Communities' technology configurations emerge as a result of leadership from the IT department, and tech stewards are supported and guided by the department. Tech stewards can have an impact on their communities and their organizations by shaping and sustaining the conversation between an organization's IT department and the community.

- **Control of resources.** An organization with an IT department both offers resources and controls technology infrastructure such as firewalls, databases or data standards, single login protocols or security standards, information management principles, and policies. IT departments often have a budget but have to balance their resources to respond to multiple demands. Stewarding in such circumstances will involve more negotiation with others than provisioning technology directly to the community.

- **Standard-setting.** Although an organization's IT department has chosen one set of cost-effective or familiar technologies and products, those technology standards may or may not suit a given community. The IT department might lump communities of practice with other groups that actually have different needs or it assumes that it knows "what's best" for all communities. This approach may conflict with the curiosity of tech stewards, who will have to limit their scanning to what is offered internally or deemed acceptable. It may also conflict with their emphasis on voluntary use and community ownership of technology choices.

- **Interplay between the organization and the community.** Technology plays a role in facilitating how a community and an organization are responsible to each other. It can allow reporting of community activity to the host organization, sharing of resources and providing access to the community across the organization, or building connections with other communities in the organization.

Stewarding across organizational boundaries

Community members who come from several different organizations often bring organization-specific technology resources and constraints. They may be using different operating systems, face diverse firewall requirements, and have different Internet access levels.

- **Finding resources and support.** Tech stewards may look to secure neutral resources, free or hosted, outside any of the members' organizations; or they may seek support from members' organizations, effectively asking them to lend technology resources to the community.

- **Bridging organizational boundaries.** Tech stewards become a focal point for identifying, coordinating, and bridging between individual members and diverse resources. Even when tech stewards seek neutral technology without an organizational mandate, whatever they choose has to work across different organizational technical settings.

- **Establishing responsibility to the community.** When tech stewards come from specific organizations, their responsibility to their community can conflict with their allegiance to their organization. Tech stewards may have to settle for the common denominator that serves the greatest number of members, rather than an ideal tool.

Stewarding outside organizations

Some communities emerge entirely outside organizations. A group of practitioners may find each other through a broader network and decide to band together more closely as a community.

- **Finding resources and support.** When stewarding outside an organization, the technology resources have to be assembled without organizational support. Tech stewards look to borrow resources from members or to use free resources. They design ways to bridge between diverse technologies, cobbling things together.

- **Defining a community space.** Communities without the ongoing support afforded by organizations may need to place greater focus on using technology to stay in touch with each other and define a space that, in turn, helps define the community. Like a community that always meets at a regular café, distributed communities need their online "place."

- **Connection to other communities.** Some free-standing communities focus exclusively on interactions within the community, while others seek to interact with broader networks and other communities. Environments such as networking sites can offer this type of connection.

Stewarding across multiple communities

There is a fourth circumstance, where tech stewards support multiple communities within, across, or outside organizational boundaries, with all the issues noted above. Tech stewards may be deeply involved with the technology supporting the communities, advocating for the needs and perspectives of a set of communities within organizations, or they may be freelance contractors who support or serve as the tech stewards for individual communities. These tech stewards can share technology knowledge and practices across communities and leverage resources and design infrastructures for communities that may not be able to create them on their own.

What about you as a technology steward?

For those of you who are or who intend to be tech stewards, take a moment to reflect about what the role means to you.

- **Are you the tech steward of your community?** Are you comfortable in this role, or are you "it" by default? If you don't fill the role, who will? Do you share the role with others? Try to join forces with others in your community who might be interested in helping with the technology steward-ship effort. Share this book with them.

- **Is your role explicit in your community?** If it is not, is it time to discuss it with your community and make it more explicit? Have conversations with members of your community, with your boss, or with the community's sponsor about what is needed in terms of technology steward-ship in your community. Stay in touch with other leaders of your community. Explore how your role as tech steward can complement theirs as you work together to support the community.

- **What skills do you have?** What skills do you want to acquire? Consider your learning needs as a tech steward. Find tech stewards in other related communities and explore how they do their work. Swap stories and share resources.

Technology stewardship for a constellation of communities

Can a PhD in physics, an interest in IT, and a varied career at a mining company land you the role of tech steward for a constellation of dozens of communities of practice? Consider Mark Bennett of Rio Tinto. His academic training, experience in various positions, and interest in technology led him to join the Knowledge Management (KM) team where he started to focus on communities of practice. When the KM team was disbanded, he continued to support a growing number of communities using the company's intranet. Though his boss knew that this occupied most of his time, it was more a personal mission than a formal function.

Technology stewardship had always been part of his community support. Peers in the industry had convinced him to acquire Sitescape as a platform for communities. Since it was not really part of the "official" infrastructure, he constantly had to advocate to the IT department for its use. He worked directly with the vendor to make sure the needs of the communities were met. His stewarding included managing the platform, training people to use it, coaching community leaders, tuning the set of features, cleaning up the site, and spending a lot of time logged in to see that every community was running smoothly. Though Mark did not really belong to any of the communities to which he provided technology stewardship, his own background made him a native in their scientific and engineering culture. He became intimately knowledgeable about each community through his ongoing watch of the site and conversations with leaders and members.

Today, with a new strategic focus on communities in the company, his leadership in supporting communities is recognized as an official job title. He still acts as a tech steward, but his support role has also taken a more visible strategic turn. He has become the voice of communities in the company. His advocacy for communities with the IT department and technology vendor was good training for this new role. He advocates to management for the strategic importance of communities with the same personal and detailed understanding of how they operate.

- **Who can help you?** You may be supported by other technologists who lend their expertise to the community. These might be people from your organization's IT department, vendors and toolmakers with a strong community perspective, friends with particular expertise, or external tech stewards who help jump-start, train, or mentor new tech stewards.

Technology stewardship is an emerging role that describes both a responsibility and a practice—an attitude as well as all the conversations, decisions, and learning that address the design and management of a community's technology infrastructure. The role has been around as long as there have been communities, but it has become more important and complex as community and technology interact more deeply. It is distinct from traditional community leadership, yet in its own way, it involves leadership in caring for the community. It includes aspects of technical support and other IT roles, yet its focus on learning in community yields a specific community-oriented perspective on technology. A growing number of people identify with the role when it's pointed out to them. They appreciate having a term to make it recognizable. Attending to the practice and developing a language to discuss it is important to tech stewards, their communities, and their organizational sponsors. Who are the tech stewards in your community? How do they discover, select, and integrate technology into the life of your community? Why did they take this role? Are their contributions recognized by the community?

Part II:
Literacy

Constructing digital habitats: community experience in technology configurations

In this chapter we refine the notion of digital habitat by proposing four perspectives on the technology involved: the tools that support community activities, the platforms in which they are bundled, the features that make them usable, and the overall configuration in which they are integrated. These perspectives help give concreteness to the construction of digital habitats as the essence of technology stewardship.

A habitat is usually defined as an area that incorporates all the environmental and biological features required for the survival and reproduction of a species (or a "community" in the ecological sense). What makes the habitat work is not just a set of physical features, but also the ways in which the species has learned to take advantage of these features for its survival. In many cases, the species itself contributes to shaping its habitat. A habitat is not fixed. As new elements are introduced, the species needs to adapt to environmental changes. A habitat, therefore, is a dynamic, mutually-defining relationship between a species and a place.

Similarly, communities of practice need habitats to learn together. These habitats have to provide the places and support the ways in which members experience togetherness. Increasingly, these community habitats include technology-based connections and places in addition to physical ones. By digital habitat we refer to the portion of a community's habitat that is enabled by a configuration of technologies. The digital habitat may be a more or less significant portion of a community's habitat, but for a growing number of communities it has become the whole habitat. What we say in this book remains valid whether the digital habitat is the full community habitat or only part of it.

Just as a natural habitat reflects the learning of the species, a digital habitat is not just a configuration of technologies, but a dynamic, mutually-defining relationship that depends on the learning of the community. It reflects the practices that members have developed to take advantage of the technology available and thus experience this technology as a "place" for a community. A digital habitat is first and foremost an experience of place enabled by technology.

Constitution of the habitat

To help make sense of the ways in which technology can be experienced as a habitat by a community we propose four perspectives:

- The **tools** that support specific community activities
- The **platforms** into which vendors and developers package tools
- The **features** that help make tools and platforms usable and "livable"
- The full **configuration** of technologies that sustains the habitat (which is rarely confined to one platform)

Constructing a digital habitat requires navigating across these four interrelated perspectives and paying attention to both the technological and practice aspects of each. Imagine that a community decides it wants an electronic newsletter tool. The members decide a blog might be good and the tech steward knows about an available blogging tool. They realize they want multiple authors, so they check the blog's features and find that it allows multiple authors. But then they realize that because the blog is on a separate platform from the discussion tool they are looking at, they would need to add another set of logins and roles to administer. They begin to worry that adding a whole new site to their configuration would confuse members. So, they look back to their existing platform, contact the vendor and find out that they can add blog functionality. In the end, the community publishes its newsletter using a blog **tool** with a **feature** that allows multiple authors. It is on the same **platform** as their discussion forums, making it easy to both administer and integrate into the community's existing **configuration**. Supporting the community's digital habitat and its evolution requires tech stewards to address all four perspectives in their interplay.

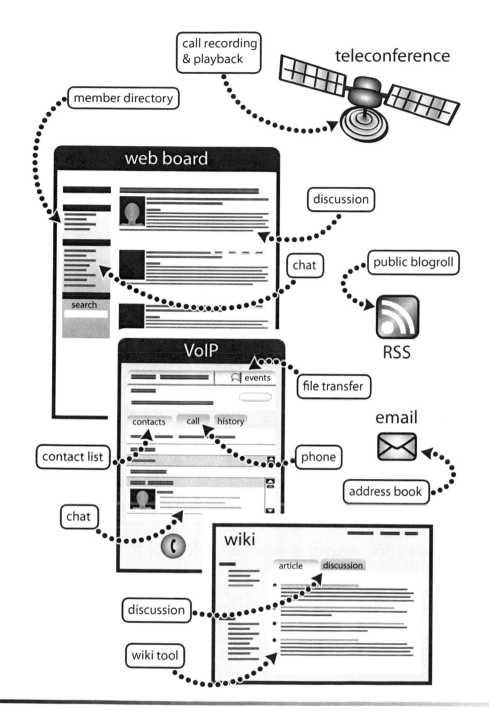

Figure 4.1 - The tools view. Tools support specific community activities

The tool perspective: habitat enabling activities

By *tool* we mean an identifiable piece of technology that supports a discrete activity in a community (for example, a discussion board that supports online conversations) or bridges different types of activities (for example, recording a phone conversation for later use).

Mapping activities to tools is not necessarily a one-to-one process. Some activities require multiple tools: for example, a meeting may require a time zone conversion website, a phone bridge, a web-based presentation, and a chat room. Conversely, many tools support multiple types of activities, such as polling, which can be used to make decisions, schedule meetings, or reflect on the community's health. Finally, tools are often used in imaginative ways, beyond their designed and intended use such as when a blog, which started out being a tool for publishing, can be used an event platform: as such, it functions as a coordination point for conversation and provides links to activities on other platforms.

Tool-focused considerations are addressed at length in Chapter 6 where we explore the complex set of activities communities engage in. Here are some questions to consider when looking at a habitat from a tool perspective:

- What is the range of activities a community engages in? Which of these activities need to be well-supported by tools?

- What kinds of activities does a given tool support? What are the designer's intentions for the tool and what other possibilities have been found? Does the tool support multiple activities? Are other tools available for use instead?

> For understanding digital habitats, the tool perspective is important because it addresses the functional adequacy of the habitat in supporting all the relevant activities. For tech stewards, the tool perspective is a primary entry point, which anchors the construction of the habitat in the specific demands of what a community is trying to do.

The platform perspective: packaged suites of tools

By *platform* we mean a technology package that integrates a number of tools available in the marketplace (for purchase or for free) that one can acquire, install, or rent. Vendors often organize a group of tools as a platform. Platforms offer communities a simple entry into using a set of tools.

While single-tool platforms do exist (such as Mailman, an open source email list program[1]), most platforms today consist of a suite of tools packaged for a defined purpose. Even previously simple applications like instant messaging have turned into complex platforms that include applications sharing, VoIP telephony, individual profile pages, personalization, directories, and search. For example, we might think of Skype[2] as a voice-over-IP (VoIP) telephony tool, but it is really many tools on one platform: it has distinct tools for one-to-one calls, conference calls, text chats, instant messages, personal and global

---

1. Mailman, www.gnu.org/software/mailman/index.html
2. Skype, www.skype.com

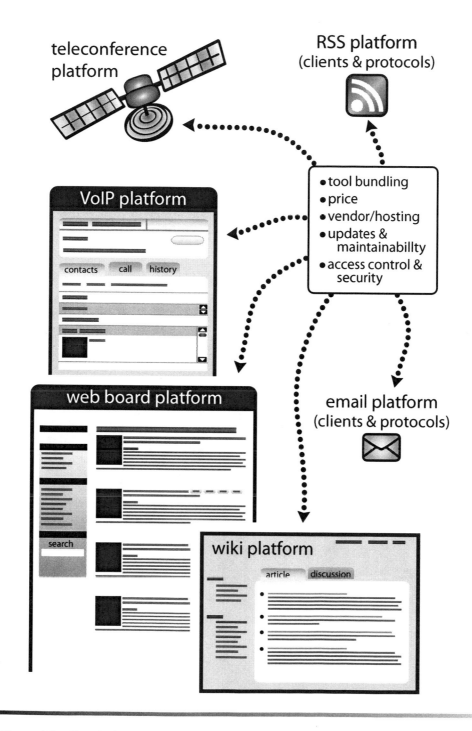

Figure 4.2 - The platform view. Vendors and developers package tools into platforms

directories, not to mention for sending files, contacts and even money. Other popular platforms package tools for synchronous meetings or blogging. A number of platforms have been designed specifically for communities and include many of the relevant tools integrated into a package (established platforms, for example, include CommunityZero by Ramius[3], eCommunity by Q2Learning[4], iCohere[5], Simplify/Ecco by Tomoye[6], but there are new entrants to the market such as Shirikisha[7], and the Idea platform by Semantix[8]).

Platform-focused considerations include questions that reflect the feasibility of acquiring, deploying, and using one or more tools together:

- How well does the platform combine the tools that a community needs? Does it provide additional features such as security or navigation that help in using each component tool?

- Apart from the usability of the individual tools, how simple (or intuitive) is the platform to use? Does the platform itself increase or decrease complexity? Can tools be turned on or off at will?

- How many members can a platform handle? How much activity is possible? Does it support multiple communities at once? If so, what are the limits and how easy would it be to launch a new community? Can subcommunities be formed easily?

- What are all the implications of gathering tools together into one platform in terms of cost, hosting, access, maintenance, and support? How can you compare all the different cost implications of alternative platforms?

- What is the relationship between the community and the platform vendor or the developer? Are they willing to modify individual tools, the platform as a whole, or allow others to do so, as the community evolves? Will the vendor develop the platform as technology evolves?

The platform perspective has to do with the building blocks of the habitat. For tech stewards it is a secondary, but important entry point, which reflects the availability of technology for constructing the habitat. To continue the metaphor of the habitat, vital nutrients for a species are rarely available in their pure form. They are contained in plants and animals in which they are bundled with other components, some also necessary for survival and some not. Given the trend towards more complex platforms, it is impossible to construct a coherent habitat without considering the ways tools are packaged.

---

3. Community Zero, www.ramius.net
4. eCommunity, www.Q2learning.com
5. iCohere, www.icohere.com
6. Tomoye, www.tomoye.com
7. Shirikisha, www.shirikisha.com
8. Semantix, www.semantix.co.uk

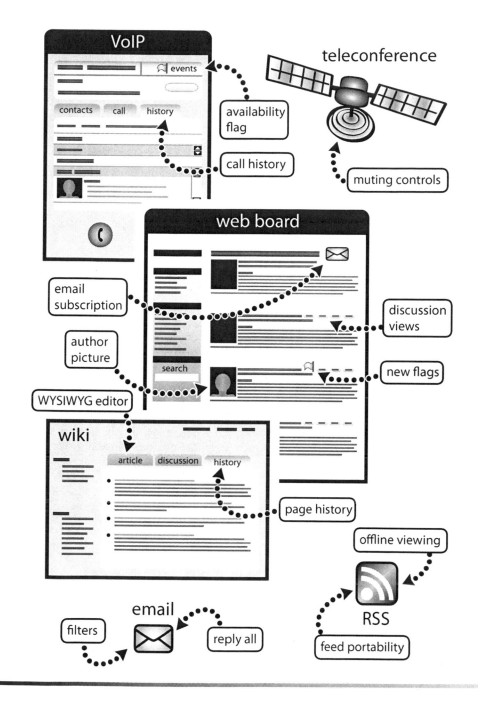

Figure 4.3 - The feature view. Features make technology usable

The feature perspective: habitat as habitability

By *feature* we mean a characteristic that makes a tool or a platform usable for a specific purpose. Some features define a tool; others add to its functionality or to the enjoyment of the experience. A phone without a microphone is not a phone, but a mute button is an element that adds functionality. Some features are very concrete facilities that are almost "mini-tools." For example, a discussion board may have facilities to preview one's own posts, statistics to see who has read a post, or "new" flags to guide people to new posts. Other features serve as subtle and pervasive design characteristics, such as a consistent user interface across a platform, consistent navigation, or adherence to data exchange or platform interface standards.

Feature-focused considerations require an understanding of how a community conducts its activities and how members get involved:

- Does the design or implementation of a feature support the specific ways in which a community conducts its activities? Are the features usable by members in practice?

- Does a feature add or reduce complexity? Can a feature be turned on or off to make the technology more useful or easy to use?

- Are key features such as menus, navigation cues, graphic elements, or controls deployed consistently and appropriately across all the tools on a platform? Will members expect certain features or recognize them because of previous experience (or from other places)?

- Does a feature inherently appeal to beginners or to more experienced users? What members' skill levels make a feature valuable?

> **Compromises for the community as a whole**
>
> A discussion tool was adopted for a global community that had members with both good and poor Internet service. The tool had the ability to send all new messages via email, enabling offline reading and response for those with poor bandwidth as well as an easy-to-read web-based interface.
>
> While the tool lacked several other features that the community leaders thought would be nice to have, the offline reading was essential to their group and therefore drove their final choice.

The features of a tool or platform determine its usability for a given community. Together the set of features determine the "habitability" of the habitat. In other words, having the right tools in the right packages is not enough: features can determine whether they are adopted or not. The devil is often in the details. To assess whether tools or platforms are a good fit for their communities, tech stewards need to look at relevant features in context. The point is not to aim for the maximum number of features but to understand how certain features meet the needs of a community or how the lack of a feature constitutes a specific problem because of the way the community operates. Individuals may have very different perspectives on features and their usefulness. It is often necessary to balance features that enhance functionality, flexibility, or security with the need for simplicity and ease-of-use.

Our "Community Tools Wiki" (described in the Preface: http://technologyforcommunities.com/tools) offers a description of tools in terms of key features and their desirability.

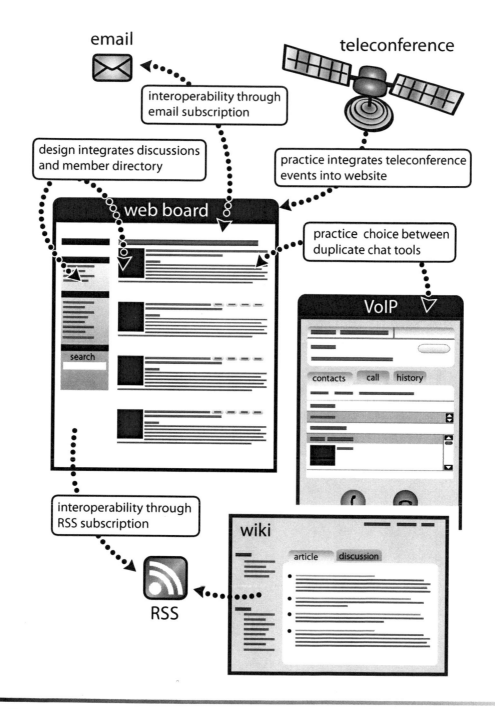

email

teleconference

interoperability through
email subscription

design integrates discussions
and member directory

practice integrates teleconference
events into website

web board

practice choice between
duplicate chat tools

VoIP

contacts   call   history

search

interoperability through
RSS subscription

wiki

article   discussion

RSS

Figure 4.4 - The configuration view. A complex configuration of technologies serves as substrate to a digital habitat

The configuration perspective: a habitat's full technology substrate

By *configuration* we mean the overall set of technologies that serve as a substrate for a community's habitat at a given point in time—whether tools belong to a single platform, to multiple platforms, or are free-standing.

For communities with complex sets of activities, the full configuration often involves multiple platforms, or selected tools from different platforms combined with a main platform. Even communities that appear to only use one platform usually depend on other tools (including backchannel emails, phone calls, public web spaces, and other means of collaboration) that are not part of the "main platform."

Exclusive focus on the community's main platform, as if it were *the* configuration, misses behaviors and opportunities that are important to the experience of the habitat. The configuration perspective includes all the relevant technologies in play, both those used "by everyone" and those brought in by individual community members, whether temporarily or permanently. This suggests that there are always two views of the configuration: the view of the community's collective configuration, and the view of the community's tools from an individual's perspective. We all have tools we use frequently and on which we depend— whether or not they are part of a community's configuration. Different members may use different parts of the configuration, depending on their purpose or role.

The evolution of a configuration

As the Knowledge Management for Development community (http://www.KM4Dev.org) has grown in size over the years, so has its technology configuration. It started out using conversations and announcements on DGroups (http://www.dgroups.org), an email list and web archive, and then added an open-source content management system (CMS), http://www.xyaraya.org.

Over time, community members found that adding and editing content on the CMS was awkward, so they added a wiki to enable members to share and co-create content more easily, particularly the summaries and FAQs derived from the conversations on their email list.

In addition, core group meetings and other synchronous conversations are held on Skype, and sometimes a member hosts a web meeting on their organization's platform. So. some bits of the configuration are always changing.

All three main community platforms have separate registrations, so it is hard for the community to fully know who is a member. The community is now thinking about what it might do next. Should they consolidate on to one platform? Build a back-end that joins all the registration databases? This next step is a big one. Stay tuned.

Configuration-level considerations move beyond those related to a packaged platform and its tools to focus on how the habitat serves the life of the community more broadly:

- Are some important functions missing in the overall configuration? Conversely, do some tools duplicate each other?

- Do subgroups evolve from using different tools for the same purpose?

- How compatible is a given platform or tool with others that the community or its members use? How tolerant are community members of the adoption of a wide variety of tools? How many technological boundaries are members willing to cross—for example, sign in to more than one web-based tool, learn to use new tools, or give up old favorites?

- Do resource limits constrain the acquisition and adoption of new tools or the use of existing ones? (For example, some people may not have time to figure out how to use what they already have.)

The notion of a configuration is important because it reflects the full complexity of the technology substrate of a community's digital habitat as a whole. Tech stewards need to view the community's configuration broadly enough to include all the relevant tools yet narrowly enough to make visualization of "the whole" practical. As configuration is defined in practice, the concept opens up questions of how communities straddle different tools or platforms, how they use them in interrelated ways. The membership or practice of some communities causes their technology configuration to be very slow-changing if not static, while other communities seem to experiment with almost every new tool that comes along, whether for an afternoon or permanently.

A practical way to start thinking about configuration is to do an inventory of your community's existing technology. Table 4.1 is an example of a table that would describe an imaginary community's configuration. (A blank table is provided in Chapter 10 to complete for your community.)

Tools can create gaps between members

A community that has its core in the US becomes focused on their teleconferences. They begin to see VoIP and telephone bridges as the center of the community's technology configuration. They have integrated the phone call schedules offered by the phone bridge service into their own personal calendar applications.

But over in Europe, they have been having detailed conversations in a web-based forum and are building up a knowledge base on a wiki. They see the center of the community's configuration as their asynchronous web-based tools. They have utilized the RSS feeds from each of these tools into their own personal aggregators, so it is easy to keep up. They become disgruntled that "the U.S. folks are not participating on the wiki," although it had been designated as the key community repository.

At a community meeting, the members discussed implications of their different views of the community's configuration.

Table 4.1 - a sample tool configuration inventory

Platform 1: Web-based discussion board

Supported Activities	← Tools →	Key Features	Usage Notes
• Ongoing community conversations • Communication of key events • Archive of community conversations • Small group workspaces	☐ Discussion tool ☐ File storage ☐ Member profiles ☐ Calendar	• Multithreaded • Has email interface • Members control their own subscriptions • Easy to create subspaces and define access	This is our primary tool and our community "memory." Most people never go to the web interface and instead use it like an email list.

Platform 2: Teleconference bridge

Supported Activities	← Tools →	Key Features	Usage Notes
• Community meetings • Small workgroup meetings • Recorded archives of meetings for those who miss a call	☐ Phone bridge ☐ Invitation scheduler ☐ Member profiles	• Members pay their own long distance • Calls can be recorded through a web interface • Can support up to 50 callers	We are using the phone bridge more and more lately because people feel they are overwhelmed with email.

Platform 3: Community Flickr account

Supported Activities	← Tools →	Key Features	Usage Notes
• Artifacts of face to face events, diagrams, and other images	☐ Photo storage ☐ Tagging/search	• Stores all community photos • Community shares one username/password	We just started using this tool, so we're not quite clear yet how it fits into our configuration.

A standalone tool: IRC Channel

Supported Activities	← Tools →	Key Features	Usage Notes
• Chat during telephone conferences	☐ Individual IRC chat clients	• IRC is a robust protocol supported by several different chat clients	We value openness and individuals are OK choosing and running own client.

Other tools and platforms...

Integration of the habitat

As the substrate of the habitat, the notion of configuration suggests pieces that "fit together." But nothing guarantees that the tools in a given configuration will be integrated in such a way that the community will experience them as a coherent habitat. Examples of important questions to consider include:

• Can members use just one sign-on to access various tools?

• Can data be passed between tools used at the same time or at different points in the community's history?

• How does the look and feel vary across the different tools? Do all the different tools appear to be located together, even if they are hosted across a series of servers and services?

- How does the navigation work between tools? Can you tell "where you are" when using any one tool and find your way "back home" to other tools?

What level of integration is required between existing tools and platforms in the configuration? How is this integration achieved and sustained as tools are added or dropped and how resource-intensive is it? How much is integration designed into the technology itself and how much does it depend on practices developed by the community in constructing its habitat?

The four perspectives described above suggest different ways in which integration can be achieved and sustained, ranging from mostly designed into the technology to mostly dependent on the community:

- Integration through platforms

- Integration through interoperability (features)

- Integration through tools

- Integration through practice (from a configuration perspective)

Integration issues of all four types contribute to the construction of the digital habitat. They are therefore important concerns for tech stewards.

Integration through platforms

A platform usually carries some assumptions about a stable, describable set of tools being of use to a community; it inherently places a value on the integration of those tools. Indeed, a good platform creates a seamless integration of tools and the possibility of linking the corresponding activities. For example, a whiteboard drawing is captured and becomes an object in the repository; an email exchange is the first draft of an entry in a wiki: ideally all objects and posts on a platform can easily refer to each other, even when they are moved from one place to another.

Underlying the integration of tools into platforms is the aspiration that the best community platform will accommodate all of a community's activities in a self-contained package. By this logic, many platforms attempt to incorporate their own version of new types of tools, as we have seen recently with blogs and wikis. This is an important trend, but there are limits to the consolidating logic of platform integration. Accumulating tools through consolidation into a platform can yield overly complex systems with much more capability than a community needs. A complex platform with many tools and features may not be responsive enough as communities recombine tools, change resources to fit their understanding of their evolving needs, and move across and between platforms to find tools with the features they need for their current purposes. As platforms try to incorporate new tools while remaining faithful to a platform's existing style, they can create versions that do not always work as expected because they are created using the platform's original underlying programming structure. So a "wiki-like" tool on a preexisting discussion platform may look like a wiki at first, but with use, does not behave in the ways that experienced wiki users are used to.

Still it can make a lot of sense for a community to seek a platform that provides designed integration among the tools it uses:

- From a practical standpoint, platform integration enables or enhances ease of use. Tools that are seamlessly integrated into a platform are likely to feel more close-at-hand and accessible since they are designed to work together. Those who are not interested in experimenting and creating an ad hoc collection of tools can rely on integration into a platform as an important way to support usability. It can be liberating not to have to worry about which tools to pick at every turn.

- From a community-development standpoint, having a central platform that integrates the community's most important tools gives the community its home and helps it assert its existence as a community. The simple action of logging in to the platform is a mark of participation and membership. Being able to see who is online, for example, creates visibility into the community and could encourage other kinds of participation, like posting or instant messages.

In such cases, the digital habitat itself becomes a source of identity for the community. It creates a boundary that delineates a separate space, access to which can define membership, but also exclude non-members. This can be either useful or constraining. But in any

---

Busy people want it all in one place

Jonathan Finkelstein and John Walber had designed and facilitated many communities, but they kept looking for a platform that integrated both traditional asynchronous tools with a fully featured web meeting tool. They recognized that communities have different needs but still want the convenience of a consolidated platform. Some communities are anchored by real-time online events and meetings. Others focus on conversation and resource-sharing in asynchronous text and voice discussion boards, galleries, and file storage areas.

Circumstances can create new integration challenges. One community they hosted was designed for 35 Mississippi counties devastated by Hurricane Katrina. This community had to use a toll-free telephone number to collect information from people who could not get online.

With the development of the LearningTimes Community platform, Walber and Finkelstein responded to their communities' strong desire for integration. Now, with the single sign on, members have a seamless navigation across a full range of community tools. In addition, they made it possible to move between any community an individual belongs to on the platform (again, with a single sign-on), providing some additional support for managing multimembership in more than one community.

http://www.learningtimes.org

case, tech stewards need to pay attention to these boundaries. Self-contained platforms may cut off the community from broader networks and from spontaneous interactions with the rest of the world. Their stark reflection of boundaries may feel claustrophobic and out of line with the networking mindset of many users of open social software.

Integration through interoperability

The importance of expandability and openness to the world, as well as the constant introduction of new tools all suggest an additional approach to integration: interoperability across tools and platforms. From a technology standpoint, interoperability is a modular form of integration that relies on open standards like XML, compatible APIs (application programming interfaces), features such as RSS feeds, or as in the box on the DITA community, a consistent navigation bar. Interoperable modules may be single tools or small clusters of neighboring tools that can be "plugged in" to each other or that can "talk" to each other. For example, calendars can be coordinated, information feeds can be combined, login and access privileges can be transferred, web services can be accessed, and applications can be "mashed" together into new functionalities.

Interoperability puts constraints on the design of technology configurations:

- Are the elements of the configuration designed with the necessary interoperability features?

- Do features that support integration across tools or platforms have the

Many websites that work "as one"

To support the DITA users group, Bob Doyle chose to use several different platforms, figuring out how to integrate them in many different ways. He used Drupal[9], Word Press[10], Media Wiki[11], Moodle[12], and Eclipse[13]—each with distinct URLs that all start with the keyword DITA. What integrates these websites is a common navigation bar with a more or less consistent set of drop-down menus, the same colors, and a standard language. Through the simple device of a common navigation bar and a naming convention, all these websites link to each other as if they were one. The community's Yahoo! Group looks different from its other platforms, but "dita-users@yahoogroups.com" is recognizable because it abides by the same naming convention. Although it takes a sophisticated tech steward to deploy all those different technologies, the approach has many advantages: separate sites increase each other's search ranking, each tool is best of breed, and the technology-literate members of the community all recognize the appropriate differences in purpose, use, and appearance between, say, a blog and a wiki.

quality and consistency that you need? Do security features, for example, conflict with tools such as RSS feeds?

- Are there features that help make content portable across tools? For example, can you export the content from one blog tool to another? Can members import content from other tools into the community using the features of the community's tools?

---

9. Drupal, www.drupal.org
10. Wordpress, www.wordpress.org
11. Media Wiki, www.mediawiki.org
12. Moodle, www.moodle.org
13. Eclipse, www.eclipse.org

Interoperability provides technical bridges between tools (and platforms) as opposed to consolidating them. It reflects the integration of a community's digital habitat in the context of a broader "digital ecosystem." With this approach, communities can extend their configuration and connections, individuals can choose the tools through which they connect to their communities, and communities can interact with other communities across their respective configurations.

Interoperability can also present challenges for tech stewards. For communities that need a high level of security, the openness to the world can create a need for constant watchfulness. In addition, expanding a configuration through interoperable tools does not necessarily yield a uniform interface. Slight differences in look and feel may throw off some members as they navigate the configuration and move to another tool. And while interoperability enables flexibility of participation, it remains a technical issue largely out of the purview of most community members.

Integration through tools

A recent proliferation of tools and services enable members to navigate across tools and platforms. Examples include tagging and RSS integrators. For instance, members of a community can tag entries on their individual blogs that are relevant to a community, in order to integrate certain aspects of their blogging with their community membership. The RSS feed of the tagged posts brings together the individual participation of members' blogs into the community view. A new breed of websites, such as iGoogle[14] and Netvibes,[15] offer integrating services that allow users to manage their participation across various platforms.

Unlike interoperability features, these tools and services put integration directly in the control of members. Individuals who belong to multiple communities, each using different platforms, will need such a bridging approach rather than platform consolidation or even mere interoperability. These individuals need personal integration capabilities by which they can coordinate and connect their various memberships, choose what to track, and decide how to participate.

When such integrative tools are widely available, a community's digital habitat supports the increasing experience of multimembership typical of today's world. It reflects the "social ecosystem" in which the community and its members live. For tech stewards, integration through tools demands attention, not only to their availability, but also to the practices associated with the use of such tools. This type of integration assumes a degree of competence and even "technological restlessness" on the part of the community—an

---

14. iGoogle, www.google.com/ig
15. Netvibes, www.netvibes.com

interest in the tools and their possibilities. This flexibility works better for communities that are technically adept and flexible, but less well for communities that lack a critical mass of early adopters or people willing to spend time hopping between tools. In some cases, those less interested in exploring new tools will feel left behind. Introducing tools and their use becomes a key task.

Integration through practice

While integration is an important technical design goal, there are limits to what we can expect directly from technology or even from the use of integrative tools. Integration also happens purely through practice. For example, the practice of producing useful notes from a face-to-face or phone meeting for publication in an online space bridges different technologies that really cannot interoperate. Even when tools are designed to work well together, members create practices that bridge between activities on these tools, for instance, a process for selecting interesting responses in a discussion to include in an FAQ (frequently asked questions) tool, or conventions for making good use of cross-referencing links between different discussion threads in an online forum. These examples of integration through practice may just evolve informally over time or be formalized in community agreements. The need for integration through practice reflects the fact that the experience of habitat is constructed by the community in its use of technology. Communities often expect their habitat to be a complete solution in a single, integrated package configuration. But in reality, even on an "integrated platform," practices to bridge across tools are necessary. Furthermore, the swiftly changing market makes the dream of a single, integrated platform difficult to deliver. Tech stewards instead have to balance between the benefits of platform integration and the emergence of new, non-integrated technologies. Helping communities develop the required integrative practices entails observing tools in use, modeling, and making practices visible—a key dimension of the evolving art of technology stewardship.

In the field of community technologies more broadly, platform consolidation, tool modularization, adherence to open standards, the emergence of integrative tools, as well as the development of integrative practices, are all simultaneously occurring trends. Vendors work to keep their products current by integrating new tools into their platforms and making their platforms compatible with other tools. They have to match the functionality available in a rapidly evolving market and avoid isolation—or they risk becoming obsolete. At the same time, the proliferation of freestanding web tools and services is prompting a quest for standards to make them universally interoperable and a rush to offer integrative tools to manage participation. This makes the notion of digital habitat increasingly dynamic and expandable, with high levels of flexibility, diversity, and individualized participation.

These trends create a dual dynamic in the construction of digital habitats, which reflect the constant interplay that exists between individuals and communities. Consolidating habitats through platform integration starts with a focus on the group around a common toolset, and uses interoperability and integrative tools to expand outward, increase personal tool choices, and open connections at the boundaries. By contrast, interoperability and integrative tools start with a focus on individual choice of tools and expand inward; out of this diversity, processes of integration enable community formation and the experience of participation. Technology stewardship is involved in this balancing act. The perspectives of tools, platforms, features, and configurations are part of the required literacy. They reveal different areas of focus for tech stewards, from the nitty-gritty of the usability of single features, to the suitability of tools for community activities, to the bundling of tools on available platforms, to the overall technology configuration. They bring additional concreteness to the notion of digital habitat as the ultimate focus of tech stewards. Indeed, helping communities construct their digital habitats—both the technology configuration and the experience of habitat through practice—is in its essence what the work of tech stewardship is about.

Making sense of the technology landscape

Here we look at the current technology marketplace as a landscape of possibilities for constructing digital habitats for communities. Asking why communities adopt technology yields two types of insights. It helps reveal what inherent challenges communities are facing for which they can use technology; and it helps articulate how specific tools act as resources for addressing these challenges in new ways. The resulting classification of tools provides a way to think about how technology not only provides functionality to support activities but also affects community dynamics.

The best way to understand how technology contributes to community life is to consider fundamental challenges communities face in trying to learn together. In this chapter, we introduce three such challenges which we believe need to become part of the "literacy" of technology stewardship. We use these challenges to make sense of the broad landscape of tools that have been adopted by communities. We end by suggesting that these challenges are useful for stewarding digital habitats and their impact on community life.

Three inherent polarities

We chose to present the key challenges that drive communities to adopt technology as pairs of concepts we call *polarities*:

- **Rhythms:** togetherness and separation

- **Interactions:** participation and reification

- **Identities:** individual and group

We use the term polarities for a number of reasons. First the notion of polarity suggests that each pole depends on the other—that considering one pole calls for consideration of the other. Second, experiencing a polarity requires a constant process of balancing between the two poles. Finally, the concept of polarity is meant to include a range of relationships and interplay between the poles—from complementarity to incompatibility, from harmony to conflict, from mutual reinforcement to tension. These polarities affect each other, but each captures a distinct dimension of the challenge of learning together.

Rhythms: togetherness and separation. Time and space present a challenge for communities. Forming a community of practice requires sustained mutual engagement

> ### Shifting for inclusion and flexibility
>
> A community of educators working to develop technology plans for their schools found it difficult to pick a time when everyone could participate in a telephone call. Rather than giving up, they held the calls with whomever could attend, and they recorded these calls. Later, those who missed a call could download it to their MP3 player and listen to it at their convenience. They shifted their participation from a fixed time to a flexible, more convenient time.
>
> Some of the educators belonged to other related communities and were having difficulty keeping up with the conversations flowing out of all of them. So, they started combining discussion board posts from multiple communities (via RSS) into one application, which allowed them to shift their participation from many spaces into one personal space. This helped them participate in multiple communities without having to "hop around" to multiple online spaces.

over time. It takes more than one transient conversation; it does not arise from merely having the same job title in different locations. It requires learning together with enough continuity and intensity of engagement that the definition of the domain, the weaving of the community, and the development of the practice become shared resources. Today the members of most communities of practice do not live or even work together.[1] So separation in time and space is a fact of life that can make the experience of community difficult. At the same time, "practicing" in different contexts is often the very reason members want to interact with each other. The diversity of these contexts is a source of richness for learning together. In this sense, separation is a resource for community togetherness. And conversely, of course, learning together is a resource for practicing in separate contexts. Togetherness and separation are in a complex interplay and their alternation is conducive

---

1. In this sense, a community of practice is different from a team, even though much of what we say about digital habitats could apply to distributed teams.

to learning. Finding a productive rhythm of togetherness and separation in space and time is a fundamental community challenge.

One appeal of technology is its variety of solutions for dealing with time and space. Technology changes, in some cases dramatically, the rhythms of togetherness and separation that are possible. It creates new "community times" that are unconstrained by schedules and time zones, and "communal spaces" that do not depend on physical location. Communities use technology to hold a meeting at a distance, to converse across time zones, to make a recording of a teleconference available, to include people who cannot be physically present, to send a request or a file, or to keep up-to-date on an interesting project.[2] In discussions of technologies that do not rely on face-to-face presence, togetherness and separation are often expressed in terms of *synchronous* or *asynchronous* tools. We will also use this distinction in this chapter.[3] But it is important to remember that the real community challenge is the rhythm of togetherness and separation more generally. This challenge applies to the use of both synchronous and asynchronous tools, as well as their alternation. How do synchronous tools contribute to a community's rhythm, both because they enable members to be together in time and because they often leave traces in the form of recordings or transcripts? In an asynchronous conversation, how often do people have to post something for the rhythm to sustain an experience of togetherness? In other words, the synchronous and asynchronous capabilities of tools are part of a much more complex story of community rhythm.

Interactions: participation and reification. The polarity of participation and reification is a process of meaning-making that is fundamental to the learning theory underlying the concept of communities of practice.[4] On the one hand, members engage directly in activities, conversations, reflections, and other forms of personal *participation* in the learning of the community. On the other hand, members produce physical and conceptual artifacts—words, tools, concepts, methods, stories, documents, links to resources, and other forms of *reification*—that reflect their shared experience and around which they organize their participation. (Literally, reification means "making into an object.") Meaningful learning in a community requires both participation and reification to be

---

2. This ability to connect any time anywhere creates a community version of what has been called "time shifting." Wikipedia, *The Free Encyclopedia*, "Time shifting," http://en.wikipedia.org/w/index.php?title=Time_shifting&oldid=212595015 (accessed July 15, 2008).

3. This reflects on the early groupware explorations of Bob Johansen, David Sibbet, and others in which they talked about the four quadrants of interaction of same time/same place, same time/different place, different time/same place, and different time/different place. Robert Johansen, D. Sibbet, S. Benson, et al., *Leading Business Teams: How Teams Can Use Technology and Group Process Tools to Enhance Performance* (Addison-Wesley, 1991).

4. For more in-depth discussion of this polarity, see Chapter 1 in: Etienne Wenger, *Communities of Practice; Learning, Meaning and Identity* (New York: Cambridge University Press, 1998).

present and in interplay. Sharing artifacts without engaging in discussions and activities around them impairs the ability to negotiate the meaning of what is being shared. Interacting without producing artifacts makes learning depend on individual interpretation and memory and can limit its depth, extent, and impact. Both participation and reification are necessary. Sometimes one process may dominate the other, or the two processes may not be well integrated. The challenge of this polarity is for communities to successfully cycle between the two.

Technology contributes to both participation and reification. It provides new ways to participate in community interactions, new ways to connect with other people and be together. It also provides new ways to reify what matters about being together—to produce, store, share, and organize documents, media files, links, and other artifacts, whether they are collectively or individually created. Technology even pushes the boundaries of both participation and reification by making it easier for a community to open up to the larger world—for instance, to decide whether to publish artifacts and invite comments publicly or to hold its work within its boundaries. Most important, technology affords new ways to combine participation and reification. For instance, augmenting a phone conversation with a web-based whiteboard supports new forms of co-authorship that casually mix conversations with written words, images, and sounds. Similarly the ability to comment on a document adds a conversational dimension to the storage of artifacts. Technology can even change how it feels to be together face-to-face, for example, by allowing a group to take notes together or edit a set of slides during a discussion.

Identities: individuals and groups. Learning together is a complex achievement that weaves communal and individual engagement, aspirations, and identities. Yet, while togetherness is a property of communities, it is experienced by individual members in their own ways. A crucial point worth repeating about learning within communities of practice is that being together does not imply, require, or produce homogeneity. Learning together often leads to disagreement and the discovery that people see the world (including technology) very differently. Disagreements and divergent views are both a challenge and a resource for a community. The individual/group polarity is a subtle, paradoxical dance if learning is going to be sustained productively. Some social trends also contribute to this polarity: increasingly, individuals are not members of only one community; they belong to a substantial number of communities, teams, and networks—active in some, less so in others. Communities cannot expect to have the full attention of their members nor can they assume that all their members have the same levels of commitment and activity, the same learning aspirations, and therefore the same needs. Conversely, members must deal with the increasing volume and complexity of their multimembership (simultaneous membership in multiple communities). They have to find meaningful participation in all these relationships while preserving a sense of their own identity across contexts.

Technology contributes to the tension between individual and community. While a tool may be designed for groups, it is largely used individually, often when one is alone. Technology also increases the complexity of the group/individual polarity. By providing varied opportunities for togetherness, it also opens the possibilities for extreme multimembership. But technology can also help manage these complexities. It can make the community visible in new ways through directories, maps of member locations, participation statistics, and graphic representations of the health of the community. It can provide tools for individuals to filter information to fit their needs, to locate others, to find connections, to know when and where important activities are taking place, and to gather the news feeds from their various communities in one place. As multimembership becomes more prevalent, tools to manage the group/individual polarity are an increasingly central contribution of technology.

These three polarities do not cover everything there is to say about the challenges of community formation and development. Nor do they account for every detail of the usefulness of technology or specific tool adoption. But they do capture something fundamental about communities—something that has to do with their physical, social, and political nature, not the technology. Being so fundamental, however, is also the reason these polarities have turned out to be a productive lens through which to look at technology from the perspective of community challenges and aspirations.

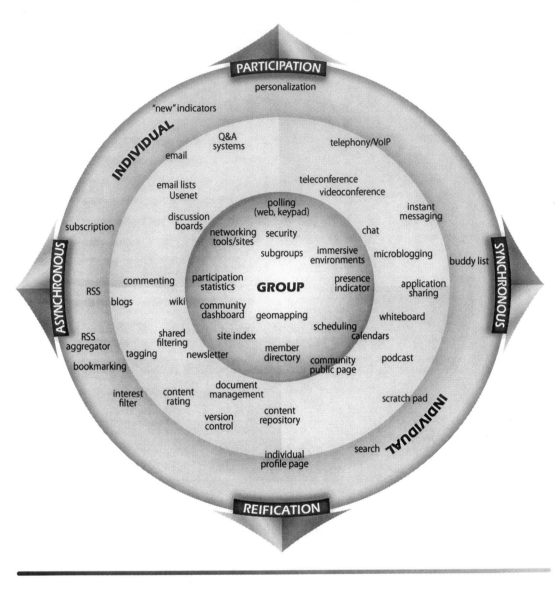

Figure 5.1 – The tools landscape

A snapshot of community tools

Using the lens of the three polarities to examine the technology landscape from a community perspective has yielded Figure 5.1. It is an attempt to capture in a single diagram how a representative sample of tools sits with respect to these polarities. This lens provides a language to explore the role of tools in our communities and to think about how certain tools tend to influence community life in one direction or another.

The three polarities define various regions in the figure:

• The horizontal dimension represents rhythms of togetherness and separation by placing synchronous tools toward the right and asynchronous tools toward the left. Again the synchronous/asynchronous distinction is not the whole story about separation and togetherness. It is merely a useful way to talk about characteristics of tools that tend to create different kinds of rhythms in communities because they enable different combinations of togetherness and separation.[5]

- The donut of the middle band represents the region of tools for interactions, with a continuum between participation and reification on the vertical axis. Toward the top are tools that support participation in conversations and activities, and toward the bottom are tools that support the creation, storing, and sharing of artifacts.

- The center circle and the outer band, respectively, represent the polarity between the group and the individual. The center circle focuses on the collective, with group and site management tools. The outer band focuses on the individual, with tools for managing participation on the part of individual members.

We have placed tools on the diagram in locations that give some insights as to a tool's most typical use with respect to the three polarities. A tool's location in the diagram should be interpreted in terms of both regions and boundaries. In other words, the location of a tool in one region is significant, but its location with respect to other regions is also significant. For instance, chat and instant messaging are both in the region of synchronous participative interactions, but chat is more toward the group and IM is more toward the individual.[6]

Many tools could appear in more than one region. For example, discussion boards can be used synchronously as the boxed story shows, but their primary intended use is asynchronous, so we have placed them in that region. Stating the design intention and actual use of a tool by a community is a matter of context and subject to debate. The diagram offers just one possible interpretation—an impressionistic, evocative overview. It should not be seen as an absolute placement or even less a prescription. Rather it is meant to generate

Unexpected use of technology through everyday inventiveness

Members of one community that crossed organizational boundaries were frustrated because some of the members were excluded from chat sessions by their organization's corporate firewall. They discovered they could meet instead on a discussion board, which wasn't blocked by the firewall. In fact, they used the discussion board, traditionally an asynchronous tool, in real time by refreshing the discussion page continually. Then they discovered an interesting advantage of this inventive use of discussion-board-as-chat facility: they could conduct a kind of *threaded chat*, discussing multiple topics in parallel by using the threaded mode of the discussion board. This approach would have been very difficult with the single stream of a conventional synchronous chat tool.

---

5. For simplicity, the diagram does not include face-to-face tools and assumes that representation in "immersive environments" through avatar-based co-presence such as Second Life is still separation in space.

6. The diagram does not encompass all possible tools, as new ones arrive daily. By focusing on these three polarities, we have not attempted to represent other important, but more technology-oriented, distinctions among tools, such as push versus pull, text versus voice, or dynamic versus static. A tech steward could also examine the community's configuration with those lenses.

useful conversations about tools and their use in communities.

Locating tools to reflect the intention behind their design on the diagram is not meant to suggest that all these tools were designed for communities. A large portion were not, but were adopted because they addressed a community need. This process of adoption can be quite creative as the vignettes in the box suggest.

The next two sections describe tools in terms of their locations in the diagram. We assume

> **Using tools that were not designed for communities**
>
> Communities of practice are skillful at putting all kinds of tools to good use, regardless of their designer's intention. A group of software developers keeps a chat room open 24 hours a day to trouble-shoot issues and socialize on a global level. Digital photographers start going on field trips together ("photo walks"), using a web-based photo sharing site to share and critique each other's photos. Neither tool was designed for a community of practice, but in each case a community has put the tool to work in order to learn together.

that you have general knowledge of the tools and their functions. A brief definition of these tools is provided in the glossary at the end of this book. For a more detailed description of these tools and their main features, as well as explanations of how they are used in communities, we invite you to visit (and contribute to) our tools wiki, described in the reader's itinerary in the introduction.[7]

The "classics" at the core of the regions

Some tools at the center of the main regions are classic, well-established community tools today: email, discussion boards, teleconferences, document repositories, user profiles, and member directories.

Telephone and teleconferencing are classic examples of technologies that were not created for communities but are now well-established tools for community interaction. Today, voice over IP (VoIP) and web-based teleconference facilities are making it increasingly easy for communities to connect through voice. Both telephony and teleconferencing are in the synchronous participative region in the diagram, but the telephone is on the individual side of the donut while teleconferencing is on the group side. Voicemail and telephone call recordings could end up on the asynchronous side, exemplifying how a tool is extended in practice.

Email, email lists, and discussion boards are other community classics: they are located squarely in the asynchronous participative region. Email is shown toward the individual region because it is used primarily for individual communication. Discussion boards are shown toward the group region because they can provide a collective experience by assembling posts in one location so that it is easier to see the whole discussion at once. Email lists are located halfway between groups and individuals. They provide much of the same

---

7. Community tools wiki, www.technologyforcommunities.com/tools

functionality as discussion boards but with a greater focus on serving individual participation by delivering posts to participants in their inboxes. Note that hybrid versions are increasingly common, allowing participants to receive emails or feeds of posts or regular digests with a link to the discussion board.

In the reification region at the bottom, providing a repository for digital artifacts (often referred to as *published content* or simply as *content*) was one of the early uses of Internet or intranet technology for communities. More specific types of documents that have become common for communities include a regular newsletter and a calendar, both shown close to the group region because they provide a view of the community. The newsletter gives an account of what is happening in the community in an asynchronous communication mode; therefore, its location is on the asynchronous side. The calendar provides an overview of synchronous events, hence its location on the synchronous side.

Also around the reification region are traditional group and individual documents: the member directory, which is intended to give an image of the whole group, and the member profile, which gives information about the individual. These two tools are on the group and individual sides of the reification region, while polling, for instance, is also in the group region but up toward the participative region, because it allows a group to explore the range of opinions in its midst. Presence indicators are at the synchronous boundary between participation and reification, located in the group region because they give a sense of who in a community is online at a given moment, while buddy lists do the same thing from the perspective of an individual (that is, which friends are online).

New generations of hybrid tools

More recently, hybrid tools have appeared that straddle the boundaries between the main regions. While the classical tools are in a continual process of refinement, new tools seem to emerge mostly at the boundaries between established regions. These new tools bridge the traditional regions to offer new ways for communities to address the polarities.

Blogs and wikis are a part of this new generation of hybrid tools. While different from each other, both combine participation and reification in innovative ways by (a) moving from a centralized to a distributed publishing model, and (b) including a participative structure around documents.

Coding the "glue"

Stephen Downes, a leader in learning and the online environment, noticed that more and more useful blogs were being written by educators, particularly those interested in the role of technology in learning.

This loosely bound "edublogger" community was not easily visible as a whole. Most often, people discovered one blog, and then followed links to others.

To make this easier, Stephen wrote some code called EDU-RSS designed to aggregate all these blog feeds. He then made the code freely available for others to aggregate their community's disparate blogs.

http://www.downes.ca/edurss02.html

Blogs were not designed for communities; they were born out of a desire for easy web page publishing. By making publishing easy, they also made it personal, adding individual voice to the conversations in cyberspace. With the addition of comments (where readers post their thoughts about specific blog posts), blogs became more conversational. While they started to resemble discussion boards, blogs retain the continuity of the blog owner's voice. This can make it more difficult to keep a sense of the continuity of the group conversation, which now is scattered across various blogs. To address this, communities have devised ways to keep a connection across their blogs by using tools to aggregate the posts of multiple blogs, linking to each other in posts and in *blogrolls* (lists of blogs one reads frequently) or using tools that aggregate blogs and other elements, such as Pageflakes[8] or Netvibes.[9]

Wikis also straddle the boundary between participation and reification. They make it possible for readers to participate as writers, creating a kind of document-focused conversation among authors. This conversation happens through their modification of the document, though most wiki platforms now include a separate discussion forum associated with each page on the wiki. Social and technical barriers may prevent people from participating in editing the wiki—for instance a reluctance to modify someone else's writing. Still, the reified product remains dynamic by always reinviting active participation.

In the diagram, the hybrid nature of blogs and wikis is reflected in their location halfway between participation and reification. While similar in this respect, wikis and blogs address the group/individual polarity in opposite directions, hence their locations on opposite sides of the diagram's donut. Wikis represent the voice of the group and the identity of the community. Open, shared editing means that the text produced by the community is the community's property, with individual contributions melding into one product. After readers stop editing, one can assume the text represents the voice of the community. By contrast, blogs emphasize the voice of authors (individuals or subgroups). Even when blogs become very interactive, the emphasis is on individual identity through the voice of the author(s).

These new tools give rise to further bridging tools. Because blogs and wikis make publication easy, they generate a huge increase in published content representing individual and community voices. This onslaught of content, in turn, creates a new challenge to make sense of all these voices. RSS, or "Really Simple Syndication," has emerged, allowing an individual to subscribe to any number of blogs and receive updates from those sites. RSS aggregators make it possible to combine and recombine published posts from various sources in ways customized to the individual. Therefore, these tools are located on the asynchronous side of the individual region, next to the boundary between participation and reification.

---

8. Pageflakes, www.pageflakes.com
9. Netvibes, www.netvibes.com

Once individuals have a way to track emergent content and ideas, they can comment on another person's blog or even use their own blog to respond to other blogs. With facilities such as *trackbacks*, where one blog alerts another that it is being referenced, individuals can be alerted to posts referencing their posts, enabling rapid replies and leading to new comments and new content. New links between texts and people emerge, bringing the individual voices of blog authors into wider networks and communities in the *blogosphere*. Authors and the text they write are linked by common domains and personal relationships, but compared with earlier tools such as email lists or web boards, have looser boundaries around a conversation, which is now shaped by the attention of readers rather than the platform of a discussion board or email list. Individuals no longer have to stay within one tool set or platform but can move more easily across tools and communities, within a set of protocols.

Social networking sites, and the technologies that drive them, provide the ability for individuals to make their identities visible to closed or open networks, and to seek connections with others—affording a basis for community formation. Many of these tools had their origins in social sites primarily focused on business or personal networking such as LinkedIn[11] and Facebook.[12] Experience has shown that finding shared affinity leads to community formation, and most of these sites now have group facilities. Flickr[13], the photo sharing site, also enables members to find and interact with other members who share an interest in a particular photo subject or practice. "Following" the additions of photos by members with a similar interest creates connections, possibly leading to the formation of a community within the larger Flickr network.

Tools offering new possibilities

Anna and Mikel Maron's first efforts to introduce a wiki as a knowledge repository for a United Nations community of practice on water governance were rejected because wikis seemed insufficiently structured. Repositories were supposed to be fixed and tightly controlled. This new tool did not fit the U.N.'s experiences of the past.

The Marons still sensed a useful possibility for the wiki: it might offer the ability for others to contribute and participate easily in the making and use of a knowledge base. So, they included a wiki in a workshop tool set, christened the WaterWiki.[10] It was intended as more than a repository: it could be a place that would allow open contribution and support meaning-making practices and learning by its users. As participants used the wiki, they not only created a community information repository, but they also demonstrated the ease of contributing and a practice of negotiation of meaning.

The wiki supported a new practice. This practice made a difference, and now wikis have been adopted as a useful tool by the Water Governance community of practice.

For more information, see Anna Maron and Mikel Maron, 2007, *A stealth transformation: introducing wikis to the U.N.*, Knowledge Management for Development Journal 3(1): 126-130

www.km4dev.org/journal

---

10. UNDP Water Wiki, http://waterwiki.net/index.php/Main_Page
11. LinkedIn, www.linkedin.com
12. Facebook, www.facebook.com
13. Flickr, www.flickr.com

Avatar-based, immersive environments, which make us feel like we are in another world (for example, massively multiplayer games like World of Warcraft[14], interactive environments like Tapped In[15] or Second Life) create new places for communities to form and interact. They offer spaces for virtual co-presence that are fertile for exploring other identities and new social worlds. These stimulate social imagination and open the possibility of new forms of togetherness. Communities are exploring the use of these environments as places for meetings, as repositories to store artifacts, and as informal social spaces to build relationships.

Feeling like we are together

The NTEN network of non-profit technologists meets every Friday on a Second Life "island." The island offers smaller non-profits a Second Life presence, a meeting place, and a connection point to other non-profits interested in using Second Life in their work to raise issues, raise funds, or do volunteer work. Not only do participants exchange information and work together, but they also have the opportunity for a playful social interaction, building bonds between both individuals and their organizations. It creates togetherness and a practical base for doing work together.

http://slurl.com/secondlife/Progressive%20 Island/217/180/25

The entire bottom-left area of the diagram has seen a lot of activity lately with the appearance of tools such as tagging, shared bookmarking, and interest filters that take advantage of social relationships to sort through the overabundance of information. Indeed, the flurry of activity in this area of the diagram reflects an increasingly sophisticated convergence of informational and social tools. For instance, the Social Source Commons[16], a site for sharing information on collaborative tools in the non-profit area, allows members to search for tools not only by functionality, but also by "who is using/recommending what." Members can express their tool expertise in their reviews, and others can use those recommendations.

Other active areas of the diagram reflect the hybrid nature of new tools. For instance, we have placed podcasting on the right side of the boundary between publishing and interacting because it bridges the traditionally synchronous use of voice and asynchronous forms of publishing and subscription. Other interesting new tools like geomapping in the group region reflect the increasing use of *mashups*. By combining, or

Figure 5.2 – A geomap of the MPD community

---

14. World of Warcraft, www.worldofwarcraft.com
15. Tapped In Community, www.tappedin.org
16. Social Source Commons. www.socialsourcecommons.org

"mashing up," traditional applications such as directories, profiles, and mapping software, geomapping enables communities to create visual maps of the geographic distribution of their members whose location in the world can be seen with an icon. Figure 5.2 shows a geomap of the MPD community described in Chapter 1. Such mashups allow a community to see itself in new ways. As standalone applications available across contexts, they also allow a person to have a single statement of identity that can be used across communities.

Technology and community dynamics

The reason the polarities introduced in this chapter are useful to classify tools is because they reflect fundamental community challenges, with or without technology. For the same reason, they provide a useful perspective to consider general issues of community development. How does a community balance these polarities as it develops over time? When does a new development, technological or otherwise, create an imbalance that needs to be addressed? Where does your community show up across these polarities and where do you think it should be?

More specifically, thinking of community dynamics in terms of these polarities is useful for tech stewards (and community leaders more generally) because the poles draw attention to each other. When one pole dominates or when technology favors one pole, it is time consider the other. For example, when introducing a tool that focuses on reification, it is useful to wonder what participative capabilities would complement this focus—for instance with conversations for authors or readers of documents. Conversely, a community that has focused for a long time on conversations may be ready for tools that make it easier to produce and share some documents. If a community tends toward groupthink and is missing a diversity of perspectives, it is time to introduce a tool that gives individuals more of a voice. This might mean adding blogs in addition to discussion forums. A community that is feeling fragmented and disconnected over time and space may benefit from a tool that can foster a greater sense of togetherness, like a web meeting tool or a presence indicator.

From a community cultivation perspective, we can picture the polarities as three sliders, like rheostats or light dimmers, that we can use to assess and manage a community's technology:

Togetherness	⬅ ? ➡	Separation
What are sources of togetherness and separation and is the rhythm optimal for the community?		

Participation	⬅ ? ➡	Reification
What kind of reification needs to be produced and how can this be integrated into forms of participation to generate meaningful learning experiences?		

Individuals	⬅ ? ➡	Groups
How much conformity of participation and tool use does your community want and how flexible does it need to be to accommodate individual choices and multimembership?		

We can slide the indicators toward one or the other end of the polarities depending on the evolutionary path of the community. Some communities intentionally exist firmly toward one end or the other of each polarity. Others need to keep an eye on the effects of tools and adjust things by moving the slider between the ends of each polarity: for instance, when is the configuration of tools excessively slanted toward togetherness in time, or toward reified artifacts like documents, or toward group processes, and what should be done to redress the balance?

Today's technology market is a kind of candy store for communities: it has an abundance of resources that can be useful to communities – to meet needs that may or may not be recognized. Thinking in terms of polarities that reflect fundamental community challenges helps makes sense of which technologies are likely to provide useful functionality for addressing these needs. Since the landscape of technology is very dynamic and new possibilities appear so frequently, the polarities help us think about how technology affects the balance of a community's habitat. The main point is that introducing a tool affects a community's digital habitat as a whole. It is not simply adding functionality; it is an intervention within a system of polarities that shape the dynamics of a community.

Chapter 5. Making sense of the technology landscape

Community orientations: activities and tools

6

People experience being part of a community in a wide variety of ways: communities have different styles. That's why different habitats work for different communities. This chapter organizes this diversity into nine distinct "orientations" we have observed in practice. Each orientation is associated with a set of tools that supports its patterns of activity. The optimal configuration for a community includes the complement of technologies that are aligned with its key orientations.

Communities learn together in different ways: some meet regularly, some converse online, some work together, some share documents, some develop deep bonds, and some are driven by a mission they serve. We say that these communities have different orientations toward the process of learning together. An orientation is a typical pattern of activities and connections through which members experience being a community. We have observed nine orientations that have implications for the selection of technology (the order is for presentation only and does not suggest a ranking):

1. Meetings

2. Open-ended conversations

3. Projects

4. Content

5. Access to expertise

6. Relationships

7. Individual participation

8. Community cultivation

9. Serving a context

These orientations reflect the importance that communities place on various ways of being together. If we say that a community is meeting-oriented, we mean that having regular meetings is a key element of how it functions as a community. Having meetings is probably not the only thing the community does, but whatever else it does, meetings are an essential part of its "DNA," so to speak.

Orientations are not mutually exclusive. For example, a community that is meeting-oriented may also keep a very comprehensive collection of community resources, making "meetings" and "content" its two primary orientations. It may also have a member directory or other technologies that support relationships. The nine orientations combine with various degrees of emphasis to create the overall style of a community. Although many communities do a bit of everything, typically some orientations dominate, giving the community a distinct feel.

The orientations of a community are not fixed: their mix changes over time as the community evolves. New orientations emerge, existing orientations change in importance or characteristics, and old ones disappear. Changes in orientations usually will have implications for the technology configuration that a community needs.

Orientations provide a framework for considering technology from the perspective of the life of a community, with a focus on what is unique about a given community. They offer a place to start thinking about how technology can support a community's critical

Artifacts or conversations?

In the very early days, the community that is now CPsquare had a strong orientation to meetings. Its members connected by email and face to face. Face-to-face events became more difficult after 9/11 due to travel restrictions, so the founders had to rethink how the community would function and decided they needed to adopt an online platform. The one they initially selected had a very rich complement of tools for community activities, but it had evolved out of a platform designed to facilitate web publishing by groups of people. This orientation, it turned out, was not well aligned with the orientation of the community. At that stage, members were not as interested in the creation of common artifacts as they were in connecting through informal conversations. The platform didn't catch on, and the community migrated to a platform that had evolved out of a discussion board and was a better fit for the orientation to ongoing, in-depth conversations.

http://cpsquare.org

Chapter 6. Community orientations: activities and tools

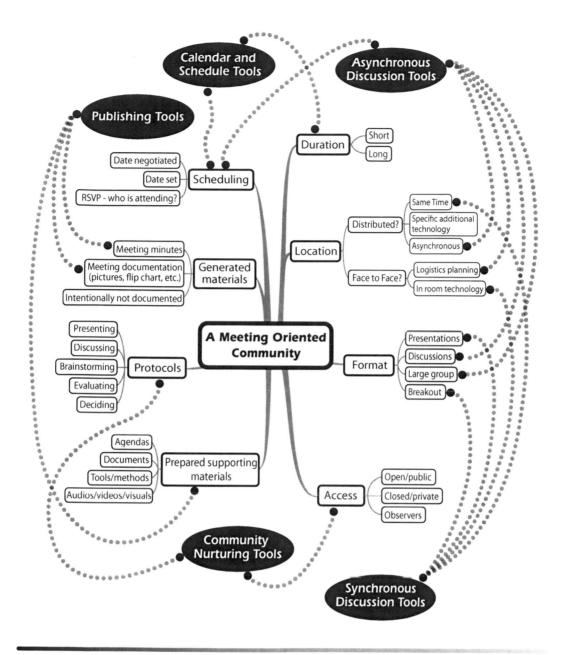

Figure 6.1: The potential configuration of technologies serving a community's orientation to meetings

activities and style. Like communities, technology platforms often do a bit of everything, but tend to focus on (or work better for) some orientations more than others. This often reflects their origin in web publishing, conversations, team support, or networking. The fit between the orientation(s) of the community and the orientation(s) of a platform is something to consider carefully. For example, Figure 6.1 gives a sense of the technology implications of a community's orientation toward meetings.

The following sections describe each of the nine orientations, with a focus on specific implications for technology. Because this chapter is long, we follow a fixed format for ease of reading and to allow you to choose where to focus:

- Each section starts with a brief *definition* of the orientation and lists the main *variants* we have seen.

- To sharpen the definition we propose some distinctive *signs of life*—indicators that the orientation is alive and well for a community.

- We list a few *success factors* we consider critical to communities with the given orientation.

- We suggest a few *questions* to consider if the orientation seems important enough to warrant configuring a set of tools to support it.

- This leads to a paragraph on the *technology implications* of the orientation.

- Included with each orientation is a table that matches a list of typical *activities* with examples of tools that can support them. A third column also includes brief "practice notes" that reflect our experience using specific configurations of tools to support a given orientation.

The lists of tools in the tables below are suggestive rather than exhaustive. You can find more details about most of the tools and some of their features in the wiki that we created to accompany the book.[1] We are not proposing a one-to-one mapping between tools and activities; many activities require more than one tool or even several *combinations* of tools. Conversely, many tools are flexible enough to be used to support several different activities. For instance, as mentioned before, polling can be used for many different purposes. Still, given the breadth of the potentially relevant tools shown in Figure 5.1 (Chapter 5), using orientations to define a required toolset is a productive exercise.

Orientation 1: Meetings

Many communities place a great emphasis on regular meetings where members engage in shared activities for a specific time. These meetings, and the visible participation of members, assert the community's existence. The main variants of this orientation include:

- *Face-to-face or blended*: People come together in one location or join a face-to-face meeting by a phone or video connection.

- *Online synchronous*: Meetings occur at the same time but from different places.

- *Online asynchronous*: Meetings occur at different times and places but with a time-limited focus.

---

1. Community tools wiki, http://technologyforcommunities.com/tools

Signs of life:

Regular, well-attended meetings, with enthusiasm to participate, connection with others, and useful outcomes.

Key success factors:

- An appropriate rhythm of meetings over time with a frequency and schedule that fit the lives of members.

- Community meeting practices (for example, agendas, facilitation, or other practices members have devised to make their meetings productive).

- Attention to the experience of individual members' participation, regardless of the medium (for example, meeting protocols that help members who are calling in on the phone feel just as present as those who are there face-to-face).

- Enough flexibility in the agenda for some spontaneous interaction and raising of issues.

Questions to consider:

- What size are the groups? Are they face-to-face, online, or a mix of the two? How are participants distributed across time zones? How might synchronous or asynchronous interactions best support the meetings?

- What are the needs of the participants to accommodate language and other individual requirements (technical or otherwise)?

From weekly broadcasts and conversations to a sense of community

Although framed as a broadcast with audience participation, the weekly Yi-Tan Tech Community Calls, or "Conversations About Change" that Jerry Michalski and Pip Coburn host, are actually the melding point for several meeting-oriented communities of practice. Jerry's community has entrepreneurs, geeks, corporate technologists and the occasional artist. Pip's has entrepreneurs, major money managers, futurists and authors.

A distinctive practice of this community is to record the calls for later podcast and to have a concurrent IRC (Internet Relay Chat) channel during the phone calls to kibitz, queue up questions, have side conversations, or contribute additional resources. A member has written an application that automatically extracts and saves any URLs that are posted in the IRC channel, as well as archiving the chats themselves. The calls are recorded and available as a podcast stream.

The calls always begin with a brief, context-setting introduction that makes the recordings self-documenting; they end with the host's phenomenal summary. The consistent participation and contribution of a group of regulars enriches and occasionally guides the weekly conversations, even though many people participate only when the topic or speaker is of particular interest to them. This shared history of learning together is most evident when a speaker isn't available for a meeting, and the group is brainstorming on a particular topic or coming up with future topics. In many cases, members of the two communities have formed lasting friendships and working relationships. The rare email flurry or wiki page annotation underscores how important vibrant meetings are for this community.

http://yi-tan.com

- What logistical preparation is required for meetings, such as scheduling, agenda development, invitations, confirming attendance, and sharing of materials?

Table 6.1 - Activities and tools for meetings

Activities	Tools	Practice Notes
Scheduling and announcements	• Shared calendar • Email • Scheduling utilities • SMS to call ad hoc meetings	When members are not in the same place, it is useful to have a calendar that can send automated reminders or can coordinate with members' calendars. Group mobile phone texting tools can geolocate members, enabling instant meeting opportunities without pre-scheduling.
Synchronous interactions	• Videoconference • Web conference, webcasting • Teleconference, VoIP • Chat room • IM	Synchronous meetings hold the attention of some people and may be their preferred communication mode. Shared visuals are good for focusing group attention. Using chat during a teleconference call is a good way to take notes publicly and makes it easier for those working in a second language or having a bad connection.
Asynchronous interactions	• Discussion boards • Wikis • Email lists	Turn-taking discussions work well in discussion boards or email lists, while a wiki can be great for building an agenda prior to a meeting.
Attendance	• Presence tools • Directories • Participant pictures	For distributed meetings, it helps to know who is present and be able to read their bios. It is also useful to have a feature that highlights who is talking.
Meeting facilitation and support	• Presentation broadcast • Application sharing • Whiteboards • Document distribution/sharing • Guided web tours • Group process tools (brainstorming, prioritizing, decision-making)	Do not assume that people are engaged and not distracted during online meetings. Offering activities that people can engage in interactively increases the sense of togetherness, for example, visiting a website or drawing on a whiteboard. Long presentations without interaction are not good community-building formats and this is even truer online when people do not see each other. For these reasons, online meetings can require design and facilitation to ensure an experience of active participation.
Enabling backchannel (private side conversations for technical, facilitation, breakouts, and content purposes)	• Chat • IM • Phone • Microblogging	In complicated, large, or high-stakes meetings, it is useful to have some people who use a backchannel communication device such as IM, chat, or an extra phone line to coordinate the technical and logistical aspects of the meeting, separate from the leader or facilitator.
Member/participant feedback and decision making	• Polls, especially instant polls • IM • "Hand raising" and related feedback tools	Stopping and taking a poll to test people's positions is a good idea when it comes to collective decision-making, but it is also useful for gaining a sense of the group without co-presence or for determining if the group is following the conversation.
Creation and distribution of shared and/or collaborative note-taking for online or face-to-face meetings	• Wikis with easy refresh • Blogs • Chat rooms • Email • Photo- and video-sharing tools[2] • Electronic whiteboards (face-to-face)	With web-based tools, minutes can be taken as a group activity and can be completed by the end of the meeting. They can even be broadcast immediately. In face-to-face meetings, blogs and wikis have been used to make the note-taking immediately available to those not present.
At-a-distance participation in a face-to-face meeting.	• Phone • Video feeds • Chat • Microblogging	Some communities use a visible device such as a balloon or a picture to give members who must join face-to-face meetings by phone a presence in the room. It's a reminder to make space for the remote member's participation.
Recording	• Audio or video recording • Podcasting/vodcasting • Photo publishing	Publishing meeting recordings and artifacts is useful for those who were there as well as for those who could not make it. Hearing it again may help someone with a different primary language or when learning a community's jargon.

---

2. For example, network diagrams created in a face-to-face workshop, then shared via digital pictures on a photo sharing site such as Flickr: www.tinyurl.com/2zq3pj

- What activities happen during the meetings? Presentations (one-to-many) or sharing of files or information, discussions (many-to-many), decision-making or prioritization, or working together on materials? Do people need access to bios or pictures to know "who is talking?"

Technology implications:

Technology both changes face-to-face meetings and makes new kinds of meetings from a distance possible and productive. All phases of meetings can use technology support—from scheduling meetings and preparation of agendas before the meeting, to sending announcements, to the interactions during the meeting, to the archiving and distribution of records after the meeting.

Using technology to overcome distance and time is not always a simple translation of familiar face-to-face meeting formats. A choice of technology has to reflect the style of the community: formal versus informal, presentation versus discussion, whole group versus breakouts. In turn, technology can impose a certain style. For instance, chat-based interactions require facilitation for turn-taking when large numbers of people are involved. Certain voice-enabled, web-conferencing systems require people to queue for turn-taking, yielding an orderly but less spontaneous conversation. With web-enabled mobile phones, groups now have the ability to create ad hoc gatherings. So a community member might be visiting a city, send a message to other members in that city, and quickly set up a face-to-face meeting.

Our experiences of face-to-face meetings don't always prepare us for the slightly different issues that come up in online meetings. In online meetings, it's hard to reproduce the way new relationships form through side conversations and impromptu interactions during breaks in offline meetings. However, once relationships begin to form online, conversations and impromptu meetings can flourish, using technologies such as email or instant messaging. Information sharing, an important part of many face-to-face meetings, is easy to do online, but it may not be a very good structuring device for online meetings when other ways of broadcasting information are available. This suggests we typically give more time to relationship building during face-to-face meetings. Communities accustomed to focused, face-to-face interaction may be disturbed by the fact that people can multitask during online meetings. Multitasking may be liberating to individuals who are less interested in the subject, but can be fragmenting for the group as a whole.

Orientation 2: Open-ended conversations

Some communities rarely or never meet. They maintain ongoing conversations as their primary vehicles for learning. Whether or not these conversations are punctuated by other activities, it is the ongoing, open-ended nature of the conversations that holds the community together. Open-ended conversations are common when a community is co-located and people keep the conversation going as they "bump" into each other. For online communities, the main variants of this orientation include:

- *Single-stream discussion*: Fairly loose discussions occur, with a spontaneous exchange of information, questions, comments, and statements of opinion—all in one thread.

- *Multi-topic conversation systems*: Distinct topics proceed in parallel, either with multiple threads in one conversation or with multiple conversations.

- *Distributed*: A combination of blog posts and comments, individual emails, microblogging, social networking sites, and instant messages are available without a central repository for all messages. For instance, conversations take place across blogs: bloggers pick up a theme from another person's blog post, and discuss that topic on their own blog, possibly leading other people to pick it up on their blogs. A common tag (keyword) used by both bloggers on their posts may tie interactions together. These interlinked strings of comments and exchanges across blogs create a sustained conversation. At any point, a new posting can reignite the conversation.

Signs of life:

A sustained flow of contributions and responses.

Open-ended conversation grows beyond a single stream

Solucient is a provider of medical software. The Solucient customer community started as a small project to test the idea that customers might be the best sources of answers for other customers if given a tool for open-ended conversation. Using a single Yahoo! Groups email list, Solucient created a customer support community around a single product. Once the first group was up and running, it was clear that the community was of interest and useful to customers, reduced customer support loads on the Solucient staff, and would be useful in supporting Solucient's other products. Additional Yahoo! Groups were started for other products.

As the early single-product communities grew, Lee LeFever, the community manager, noticed how much people wanted to talk to each other about ideas in their work, beyond their use of Solucient's software products. There were other conversations they wanted to have with each other once they connected around the support discussions. This created difficulties with the collection of different email groups they were using, defined by software products. The community needed more flexibility for creating conversations on various topics that may or may not be related to product support. It was time for a move to a platform that would enable them to scale up, segment subcommunities for the different products and topics, and allow everyone to choose which topics they wanted to join independently of the particular Solucient product they were using.

To support a community of rich conversations of different types, Lee and his team decided that a conversation-oriented commercial community platform would allow them to move quickly, reducing in-house development time and providing a range of needed functionality. They selected Web Crossing and then carefully planned a transition from the Yahoo! Groups that took into account both the technical and community issues. After they successfully migrated most of the members to the new platform, the community entered a new phase of development—into a more complex system of open-ended conversations that enabled it to continue to grow and thrive.

http://www.solucient.com/clientaccess/clinical.shtml

Chapter 6. Community orientations: activities and tools

Table 6.2 - Activities and tools for open-ended conversations		
Activities	**Tools**	**Practice Notes**
One-topic-at-a-time conversations	• Email • Email lists • Chat • The comment feature of blogs • Group mobile phone messaging (SMS)	As illustrated by the Solucient story, a community's early conversations may be focused in a single topic but may grow more complex over time, suggesting a shift to multiple parallel conversation options.
Multiple concurrent topics of conversation	• Web-based discussion boards • Wikis • Blog discussion tracking, categories, trackbacks, pings and aggregation services • Microblogging • SMS/text	There is a tension between keeping everyone together and allowing topics to branch out. Separate topics make for in-depth, focused conversations. But when there are too many distinct discussions for different topics, fragmentation is the risk and members can feel lost, not knowing where to post or pay attention.
Highlighting key learning	• "Frequently Asked Questions" (FAQ) area • Wikis for summaries • Tags, categories • "Thumbs up" and other rating mechanisms to mark the value of an individual post • Tools that move active discussions into primary view (for example, a "What's hot?" section on the home page)	Tools to highlight key learnings vary in the amount of intentional cultivation they require. Some are distributed and automatically aggregated, such as tags and ratings. Some require substantial attention such as adding polls to surface community feedback or creating and stewarding wikis for summaries and FAQs.
Subgroups/privacy	• Access control (who can participate) • Mechanism for reporting back to the larger group	It is good to find out early what a community's perspective is about openness or privacy. Some communities need more privacy at first, and then open up later. Many keep both a public face and private spaces. The same is true for subgroups, as fully private subgroups tend to fragment the community.
Translation between languages	• Parallel discussions for manual translation • Automatic translator window • Automated translators integrated in discussions	As more communities operate globally, issues of language and meaning-making across languages are becoming more important. Automated translation technology is still not quite at the stage of supporting easy communication.
Archiving	• Web-based repositories for email lists • Automatic archiving in discussion boards • Permalinks in blogs • Tag clouds	Cleaning up is very important. Communities often want to hold on to old discussions. Ask the question: Will anyone ever really look at them? Do dormant discussions make it hard to find current, active conversations?

Key success factors:

• Enough variance in topics to keep it interesting but not so much as to create subdivision into separate communities.

• Enough contributions to feel active, but not so many that members get overwhelmed.

• Active participation by a representative segment of the community. (This does not mean everyone. Online open-ended conversations typically involve a large number of readers, or *lurkers* as they are sometimes called. But it is important to make sure that the conversation is not hijacked by a small vocal group whose interests do not reflect the whole community.)

- Well-organized conversation archives that avoid circular conversations and help newcomers get up to speed.

Questions to consider:

- Do your members want (and have enough commitment) to engage with each other on an ongoing basis?

- Are conversations focused on one topic and/or over a specific time frame, or do they branch and evolve over time?

- Does everyone in the group have to have access to all conversations? Is there a need for private conversation? What is the role of backchannel (private) conversations in the community's public conversations?

- Do conversations need to be harvested, "captured," or archived for easy access in the future?

- Is the community multilingual? Are there translation needs? Do different language conversations happen in one area or in separate areas?

Technology implications:

Email lists and chat rooms work well for single conversation streams because the conversations all happen in one place, with the primary focus on responding to the most recent entry. But as the conversation moves on, topics typically get dropped. It is difficult to deepen the conversation into multiple topics in parallel without adopting more sophisticated practices that use threading or email filters. Tools that allow parallel streams of conversations are inherently more complex to use because each topic develops its own context, and contributions need to be made in "the right place." Traditionally, web-based discussion forums have been used for parallel conversations. Newer tools such as blogs and wikis are useful for single-topic streams, for instance through the use of comments, but they can also work for parallel conversations with RSS feeds, categories, and tags. Pairing discussion tools with polls and wikis can help make them useful for group processes and knowledge retention. For example, the KM4Development community has a wiki separate from its mailing list where members are asked to summarize key discussion threads they initiated on the community mailing list.[3]

---

3. Knowledge Management for Development Wiki, www.km4dev.org/wiki

Orientation 3: Projects

In some communities members want to focus on particular topics, go deep, and collaborate on projects to solve problems or produce useful artifacts. Learning is not just a matter of sharing knowledge or discussing issues. Members need to do things together in order to develop their practice. Projects usually involve a subgroup within the community: participating in the project team on behalf of the community becomes an important connection to the overall community. The main variants of this orientation include:

- *Co-authoring*: Documents and other artifacts are produced collaboratively.

- *Practice groups*: Temporary or longer-lasting subgroups focus on an area of interest, usually with the idea of reporting back to the larger community.

- *Project teams*: Temporary teams are formed to answer a question or accomplish a specific task on behalf of the larger community.

- *Instruction*: Structured learning activities, including training and formal practice transfer, are undertaken for internal or external audiences.

Beyond conversations to projects

An informal community of online community facilitators decided they needed to learn about new online community tools. The tools were proliferating faster than any one member could keep up with, and through their conversations, they realized they had different perceptions and experiences with the tools. One member suggested organizing a series of virtual field trips to different community platforms so they could learn about them together. A subgroup convened to organize the trips, captured notes, and created summaries of each trip that they posted back to the email discussion list with a copy stored in the community's file repository. This "field trip project" sat "alongside" the community's ongoing conversations. It became a regular practice of the community that lasted a couple of years. The community built its knowledge, and the subgroup of field-trippers got to know each other better and built new relationships. The community itself became known as a place to learn more about community tools from a community facilitator's perspective.

Signs of life:

Committed engagement, as a whole or in subgroups, in producing some change in the community members' world, such as developing a useful artifact, addressing a recurring problem, or responding to a challenge.

Key success factors:

- Collective definition of projects related to the community's domain

- Coordination and leadership

- Adequate communication between subgroups and the rest of the community

- If inside organizations, alignment with internal project management process and procedures

Table 6.3 - Activities and tools for projects		
Activities	**Tools**	**Practice Notes**
Creating content together (co-authoring, collaborative writing, editing, and so on)	• Wikis • Application sharing (synchronous) • Track changes in word processors • File sharing • Workflow	It is useful to understand how much control over content creation is needed. For close tracking and editorial control, consider tools with file check in/out and version control. For more informal situations, wikis and sharing word processing documents are more appropriate.
Subgroups	Tools with features that allow: • Access control (who can participate and in what way) • Subspaces to be set up on the fly as needs emerge • A mechanism for reporting back to the larger group • Group private messaging (web or mobile phone)	Small communities can organize subgroups without a lot of technology support, but the ability to create new groups and access permissions with tools will save time in a larger community.
Project management	• Team and project-management tools (Gantt charts, timelines, task trackers, schedulers) • Calendar • Project dashboard	The voluntary nature of communities may encourage or discourage the use of specialized project management tools. For those with simpler needs, there are many creative ways of using common tools, like a simple shared calendar with project milestones.
Instruction	• E-learning platforms • Participation tracking/ completion tracking • Screen sharing • Web meeting tools	With content abundant on the web, instruction often entails focusing the attention of learners, organizing the content, and making it meaningful.
Communicating with or engaging the rest of the community or a wider audience	• Project blogs • Wikis • Screencasts	Blogs are good communication devices for getting comments, while wikis enable the audience to become active in the shaping of the product.

Questions to consider:

- Do members feel a need to "do" things together in order to learn?

- How formal and/or ad hoc is project definition and management? Do teams require private spaces?

- What are the requirements to support the collaborative activities? Coordination? Creation of artifacts? Project management? Meetings? File repositories?

- Are other members likely to want to be informed of the progress of subgroups or to become peripherally engaged in their work? What is the process for reporting out?

- What kinds of products or outputs are likely to be created, and what has to happen to the outputs?

- Is structured instruction or practice transfer part of the work of the community?

Technology implications:

Close collaboration often requires separate spaces where a subgroup can work together without being disturbed by others. For example, having contributions from outside the subgroup can be disruptive, but having non-group members see what is posted by the subgroup is acceptable. Collaboration may require common structures to work on shared artifacts, coordinate participation in precise ways, and manage tasks, particularly in larger groups. Collaborators may need tools to co-edit or create documents, calendar tools to coordinate activities, and project-management tools to track interdependent tasks. In addition, a subgroup focused on a project will often need to communicate with the community at large. Tools such as blogs and wikis that invite participation around published documents can be used to update and involve the rest of the community. Group size matters in tool selection, as some tools are more useful to small groups and some to larger groups. Some communities may want members to be able to create new project spaces on the fly, while others may want to have a more formal set-up process.

Orientation 4: Content

Some communities are primarily interested in creating, sharing, and providing access to documents, tools, and other content. Valuable and well-organized content is a useful resource for members; it also attracts new members and makes it possible to offer a community's expertise to others.

Creating reified "stuff" can be a byproduct of participation in community activities (for example, notes from a meeting) or engagement in practice (for example, sharing a template one has built). It can be a goal of participation in itself (for example, creating a graphic representation of a good practice together). Note that activities just described under Orientation 3: Projects often produce content, so that co-authoring and related tools are covered there. The main variants of an orientation to content include:

- *Library*: Providing an organized set of documents of any format

- *Structured self-publishing*: Members contribute structured objects, with consistent formats and meta-data fields (for example, book, paper, website address, personal information).

Building a library by tagging

With the advent of the Web 2.0 era, a community of tech stewards needed to understand what tools were available and useful to their communities. While they were good at scanning for the latest news on technologies, they were struggling to "bring back," organize, and use the information they were finding individually. The content was critical to their work, but it wasn't available to all of them. As an alternative to writing reviews on their base platform, they agreed instead to start tagging key sites using their http://delicious.com accounts, and then aggregate the tagged resources back in their platform automatically. Over time they agreed on a few key tags. This provided the background material for their technology planning conversations in their respective communities—collected by many, but easily aggregated and organized into one place.

- *Open self-publishing*: Members contribute any file, text, or digitized material to share.

- *Content integration*: Integrating feeds and links from various internal and external sources for organized access.

Table 6.4 - Activities and tools for content		
Activities	**Tools**	**Practice Notes**
Uploading and sharing document files	• Separate document repositories • Attachments to discussions	Many discussion boards allow the attachment of files to posts, but it may be hard to find documents later if they are embedded as attachments.
Commenting on, annotating, and discussing content	• Discussion forums • Wikis for annotation • Blogs with comment features • Web page annotation tools	Discussion may be critical for the community to "own" and fully utilize some content. Linking the two is important if files are stored separate from discussions.
Publishing self-generated content	• File sharing • Blogs • Web pages • Wikis • Screencasts	There are so many places members can publish. Their work may end up being out of view of the community. Finding ways to link external member publishing to the community is useful.
Publishing structured objects	• Content management systems • Meta-data features • Adherence to documentation standards like the "Dublin core"	When integration of diverse data resources is an issue, predefined object structures with meta-data force contributors to indicate how their contribution fits in the overall taxonomy.
Centralized editorial control (for example, organizing, approving, editing)	• Editor functions to show changes, version control • Manual editing and approval for public posting • Access controls • Workflow for routing material	Centralized editorial control makes for cleaner repositories but requires a lot of work by an editorial staff.
Distributed editorial capabilities	• Tagging • Rating • Commenting	Balance control with making it easy for members to participate in the processes.
Rating contributions	• Rating mechanism • Activity tracking • Metrics and reporting • Tagging	Balance vetting content against the need to encourage contributions from shy authors. Make sure that the use of ratings and metrics for content is aligned with your community's local culture.
Accessing internal and external content	• Search engines • Tagging tools • Subscriptions/alerts • Aggregators and newsreaders with features such as RSS, trackbacks, and pinging • Subscription links to paid content • Web-enabled mobile phones	Balance internal and external information sources to maintain focus on your community's domain and practice, while still providing the benefit of what is happening outside.
Archiving	• Time-sensitive notices • Automated archiving	Deleting or marking content as "out of date" focuses attention on current content.

Chapter 6. Community orientations: activities and tools

Signs of life:

The regular creation or identification of new material and frequent downloads or use of existing material; active involvement with content—commenting, discussing, tagging, remixing, reorganizing, and exploring relevance.

Key success factors:

- Careful and ongoing organization of content that reflects the community's view of its domain

- A flexible taxonomy that allows for growth and evolution

- Ease of publishing internally in the community or out to a larger public

- Ease of creating new content, especially in collaboration with other members

- Archiving of aging material

- The use of tools that invite active involvement with documents

- Excellent search capabilities

Questions to consider:

- How frequently are documents, tools, and other artifacts collected, created, or used in the community?

- What does the community do with the content? Is it annotated, organized, and filed, or is it constantly in flux and in use? Is there an editorial process around it? Are discussions and critiques organized around the content?

- What types of artifacts (for example, tools, reports, transcripts, or recordings) do community members need to share? How large is the collection likely to become?

- Who is responsible for organizing and archiving material? What are their needs?

- Who has access? Does the content need to be password protected or is it something of broader interest that should be accessible to public search engines?

Technology implications:

A large volume of documents and other artifacts suggests the need for technology that focuses on content management: uploading, organizing, combining, search, application of taxonomies, and editorial functions. Documents are easier to find if stored in some sort of electronic *folder*, tagged or organized under defined categories, and searchable down to the text level. But documents often derive their value in the context of interactions—pointing to the relevance of tools for conversations, comments, ratings, and tracking downloads. Communities have to balance the need to manage documents in and of themselves with the need to allow for their use in context. Beyond traditional content management systems, web technology affords new ways for communities to handle the management of its documents. There is still a place for centralized, structured organization of a repository, but the web also offers possibilities for members to engage actively with documents in a less structured, distributed fashion—whether in the collective production of documents through tools like wikis, or in the collective development of emergent structures for organizing resources through links, tagging, and comments, for example.

Orientation 5: Access to expertise

Some communities create value by providing focused and timely access to expertise in the community's domain, whether internally or externally. Communities with this orientation focus on answering questions, fulfilling requests for advice, or engaging in collaborative, just-in-time problem solving. Some even have an informal or formal research function to respond to requests. The relevant expertise may be held by the whole group or a smaller set of experts. A community may serve a larger organization or a network as a "center of excellence," with a focus on identified expertise, or may serve more informally as a connection point to access the knowledge of its members. The main variants of this orientation include:

- *Access via questions and requests*: A question or request is broadcast or directed to potential respondents; responses are often kept for future reuse.

- *Direct access to explicitly designated experts*: Experts are made available through visits by guests, consulting a center of excellence, and "following an expert."[4]

- *Shared problem solving*: A group of members is called on to help an individual solve a problem in real time.

- *Knowledge validation*: Responses or artifacts are routed to respected members so that they are fully vetted.

- *Apprenticeship and mentoring*: Learning takes place through observation of or apprenticeship with a skilled practitioner.

Just-in-time expertise

A community of sales people in a technical company has an established IM channel. When they are at a customer's site and need an answer to a question or challenge posed by the client, they can send a request to the IM channel and any member of the community who is online can respond. Sometimes, if the question is difficult, they will organize a phone conversation to discuss the issue. For them, belonging to the community means that they have immediate access to the right kind of expertise just when they need it.

---

4. This borrows from Brown and Dugid's concept of "stolen knowledge" (http://www2.parc.com/ops/members/brown/papers/stolenknow.html, 1992 Educational Technology Publications) and exemplifies Efimova's observations about blogs as a channel for "distributed apprenticeship." Lilia Efimova, "Legitimised theft: distributed apprenticeship in weblog networks," Mathemagenic, http://blog.mathemagenic.com/2004/05/14.html

Table 6.5 - Activities and tools for access to expertise		
Activities	**Tools**	**Practice Notes**
Questions and answers	• General tools such as email, chats, text messages, email lists, or discussion boards • Specialized tools such as Q&A systems, FAQ tools that compile questions and answers, or answer mining	Scale is a key issue here because of the traffic that requests can generate. High traffic requires routing of requests so they reach only likely respondents and the ability to archive questions and answers so they do not get repeated.
Expertise locating	• Member directories • "Yellow pages" tools for self-declaration of expertise • Expert ranking and/or rating • Social networking tools	By using technology to make expertise more visible, you alter community structure. Sorting out who's the expert is a significant community contribution. A community can become a filter to use expert time only when really needed. Without norms and agreements about how accessible "experts" want to be, many will avoid declaring their specialization too publicly.
Validating or rating responses and escalating questions not yet answered or with inadequate answers	• Rating tools for responses • Commenting tools • Visibly linking authors to contributions • Polls • Wikis for adding to base knowledge • Automatic routing of contributions to expert panel	Rating responses is useful in helping decide where to route questions of the same type. But it can have both positive and negative effects. Setting things up so that comments or criticisms of a response aren't seen as personal attacks can be helpful. Alternatively, giving people credit for their answers can make it more attractive to contribute.
Shared problem solving	• IM/chat or telephone • Video feed • Application sharing • Whiteboards • Teleconferencing • Discussion boards	These shared sessions often take the form of a project. See comments under Orientation 3: Projects.
Following an expert	• Blogs • Subscriptions, RSS • "Watch this member" feature • Microblogging	Traditionally following an expert meant knowing what they posted or downloaded. Now we have many other options. For example, following an expert via Twitter requires a commitment by the expert to post short, frequent messages. This can enable many to follow.

Signs of life:

Rapid and reliable responses to requests for expert advice and for specialized assistance; well-established methods of eliciting community expertise; people know who they should go to for specific expertise.

Key success factors:

• Holders of expertise known or designated (by reputation, specialty, or job)

• Quick access to reliable sources of information and/or quick response from experts

• Accurate routing to the best potential sources of help

• Reliability of responses established either by the reputation of respondents or through explicit validation processes

Questions to consider:

- Do members of your community need to get rapid access to information and advice? From each other and/or from designated experts?

- How important is the formal validation of knowledge for the community?

- How do members become aware of each other's knowledge? Are members willing to "declare" their expertise on a topic? Do people care about building a personal reputation, or would they rather not accentuate differences in skills, levels, and quality of contribution?

- Does the community serve as a center of excellence for a larger group? How should access be provided?

- Does the community regularly bring in outside experts? How familiar are those experts with the tools used in the community? What support do they need?

- How big is the pool of people who need to interact? Smaller groups can manage informally with little support, but large groups benefit from tools to help automate some processes.

Technology implications:

Common communication tools such as email, the phone, or IM can be used for questions and answers, but their use assumes that the requester knows the best source of information to contact, and they tend to limit interactions to just a few people. Such simple tools may not scale up, partly because they don't provide for the reuse of questions or answers. Asynchronous discussion boards involve more people and therefore can yield more reliable responses, but they may not work for rapid responses and can overwhelm members with traffic. More sophisticated applications that enable quick and efficient access to expertise such as expertise locators and Q&A systems are also available. These can route requests, build and access a repository of questions and answers ("Frequently Asked Questions" or FAQ), and keep track of the ratings of responses that various experts receive. Contact management and social network analysis tools can be used to map the expertise in one or more communities.

Orientation 6: Relationships

Some communities focus on relationship building among members as the basis for both ongoing learning and being available to each other. This orientation emphasizes the interpersonal aspect of learning together. Communities with this orientation place a high value on knowing each other personally. They emphasize networking, trust building, and mutual discovery. Members care about who is in the community. Sometimes this focus on relationships is purely internal. Sometimes it extends outside to connecting with others and even recruiting new members on the basis of personal connections. The main variants of this orientation include:

- *Connecting*: Networking with people with whom one is likely to find a mutual connection.

- *Knowing about people*: Getting to know each other at professional and personal levels.

- *Interacting informally*: Interacting with other community members one-on-one or in small groups.

Signs of life:

Networking, bonding, friendship, references to personal lives in conversations

Key success factors:

- Ways for people to get to know each other and build their identities.

- Opportunities to connect informally beyond participation in organized community events.

- Networkers acting as connectors with other people.

- Having individual control over personal exposure and disclosure (see the next orientation).

Questions to consider:

- Are members drawn to the community for the opportunity to connect with people as much as to find information or gain skills?

- How dependent is the ability to learn together on the level of trust and depth of interpersonal relationships?

- How curious are members about others and how willing are they to disclose information about themselves? Are members interested in investing the time and effort to build relationships and get to know each other beyond the domain-oriented interactions of the community?

- How large is the community and how widely do people need to build relationships across the community? (Complexity of creating and maintaining relationships grows with community size.) How open or closed is the community?

Beer offline and online

A group of technologists concerned with the application of an open-source content management system have a complex practice for sharing improvements to the base code, and a well-developed online information repository. Despite their technology adeptness, they realized they wanted regular, sponsored, face-to-face workshops to train others on the use of the software. But a hallmark of the community is that their gatherings always include a social element—going out for a meal or a beer—as part of their practice. They make sure people get to meet each other before the workshops start, and they often connect with each other using other social networks outside their main platform. They see themselves as connected deeply around their domain, but, more than that, as friends and colleagues as well.

In an updated version of the same process, a community of educational game designers is using Second Life as a social gathering space.

Table 6.6 - Activities and tools for relationships		
Activities	**Tools**	**Practice Notes**
Networking, finding others, revealing our relationship to others	• "Light" member directories (contact, but minimal personal information) • Social networking tools • Social network analysis tools	A key issue is the balance between access to tools that help find others and explore relationships and the culture of the community with respect to privacy and personal information.
Discovering information about others, expressing personal identity	• "Heavy" member directories (with lots of information about members) • Profiles and personal web pages • Member pictures associated with each contribution to conversations or repository • Photo gallery, photo sharing • Lists of favorites (URLs, books, songs) • Blogs	Our identities may now be shared in bits and pieces across the Internet and within diverse communities. Consider how you can tap into those sources and not ask members to "recreate" their identities solely within the community.
Knowing who is around the community and interacting informally with other individuals	• Community-specific presence indicators • Invitation to instant chat • IM buddy lists • Email • Phone, VoIP • Immersive avatar-based environments • Microblogging • SMS	Knowing who is logged in to a website or online can be useful, but being able to say "hello" or interact informally can really make the experience of the community more personal. Informal two-way communication is a method of achieving learning in a community.
Forming casual or ad hoc subgroups	• Access lists • Delegation of rights needed to set up subspaces • Geolocating tools on web-enabled phones	Some people need private space to develop trust. Others will advocate for openness. These diverse needs may cause tension in both the tool choices and the community practices.
Following others	• "Watch this member" features • Tagging • Seeing what someone reads or posts • Social networking sites • Microblogging (i.e. Twitter) • Friend aggregators (i.e. FriendFeed)	Some members may express concern about "too much availability" and interruption. Consider whether your tool allows people to control who can see they are online.

Technology implications:

Relationships are between people; therefore, technology may seem less relevant. Yet technology has turned out to provide many ways to create, sustain, and represent human connections. The web has recently seen an explosion of tools oriented toward building and visualizing relationships, particularly social networking for finding and explicitly stating relationships with other people, and social network analysis tools for representation of network connections. Some of these tools are suited to communities; some are more oriented to general networking but may be used in the context of communities.

We are often asked by people who have never seen it happen whether real relationships can develop without face-to-face interaction. In our experience they can and do develop, both in purely online settings and in combination with face-to-face. As people become more experienced in using technology, new mixtures will become commonplace. Communities are experimenting with techniques for including remote participants in face-to-face gatherings. Finding the right mix of face-to-face interaction with the many tools that exist is both subtle and challenging. The different sense of presence and even of identity that we have in immersive environments like Second Life, offer new opportunities for new kinds of relationships. In the end, however, there are no guarantees in developing relationships, even in face-to-face settings.

An orientation to relationships does not necessarily apply to an entire community at once. People often discover others in a community with whom they would like to pursue a special connection, either around a topic or an activity, or at a purely interpersonal level. A relationship orientation requires the ability to let members form smaller groups by segmenting the space with a mix of public and private subspaces. This places a premium on the ability to create subspaces easily and to distribute the ability to control access or open up these areas. Relationships also may extend outside the community, allowing a community to tap other tools, such as members' bookmarking accounts, and pull those feeds into the community without asking the member to do any additional work. Over time, new communities are emerging out of interactions on microblogging tools like Twitter.

Orientation 7: Individual participation

Learning together happens in the context of a group, but it is realized in the experience of individuals. Learning together does not imply homogeneity of learning. People bring different backgrounds, communication styles, and aspirations to their participation in a community. Increasingly, their participation in any community takes place in the context of multimembership in many other communities—a factor that is bound to give them a unique perspective in any given community or facet of community life. As a result, members of the same community participate in different ways; they have different purposes, they engage with different frequencies and different

Accommodating participation styles

A global community of practice noticed that some members strongly preferred asynchronous discussions and others preferred phone calls. The teleconferences presented time zone issues as well as comprehension challenges for those who were not participating in their first language. The asynchronous discussions posed problems for people who never found time to read them.

The diverse preferences were causing a split in the community. Both options were always available, but there was not a lot of cross-pollination across the two. So, they created a practice of posting annotated notes from the teleconferences in the discussion board area, and tried to include at least one member from the discussions in the call to bring the perspectives of one subgroup to the other.

Table 6.7 - Activities and tools for individual participation		
Activities	**Tools**	**Practice Notes**
Individualized website navigation across successive visits	• Individualized indicators of new material (for example, pointing to what new materials on a website one has not seen) • Notepads to keep individual notes or journals • Individual message center to bookmark contributions of interest	Navigation can quickly become an issue as a community site grows through member contributions. Practices for tracking one's participation and content vary greatly, and are often dependent on the member's technology skills. You can expect to see variation in this area and may find that some members play the role of "finder, filterer, and sorter" for the larger community.
Customization	• Filters (what to see and what to hide) • Individualized site maps, pointers to relevant areas, and taxonomies • Profiles (time zone, connection speed, language) • Preferences (display, look and feel, home page) • Customized search (from preferences, history, profile, or relationships) • Multi-language interfaces and translation capabilities • Choices of platform to receive content (web, email, phone, etc.) • Tagging	The more customization options offered, the more technical orientation, training, and support is needed. Consider whether enough members will benefit from additional features.
Subscriptions	• Subscriptions flagged on a website • Email alerts • RSS • Individualized digests • Alert mechanisms • Multiple routing options (email, SMS)	Subscriptions and alerts are important. Don't expect members to visit the community site regularly unless this participation is part of an explicit commitment.
Bridging interaction modes	• Recordings and podcasts • Real-time notes published through blogs or wikis • Video feeds • Informal interactions with IM, microblogging	What is important here is that people can choose the interaction mode they prefer if the modes are well-integrated.
Managing individual participation publicly	• Bulletin boards to announce individual circumstances like absences or periods of limited access • Listings of communication preferences	This is more of an issue for smaller, relationship-oriented communities where the impact of each individual's participation is important to the community's health.
Managing one's privacy	• Features of IM tools that allow members to turn on or off their availability in IM/presence indicators • Interaction tools that do not keep records or transcripts that can be accessed and viewed later • Portability of one's content across platforms	In our experience, until people realize their privacy can be managed, they may be shy or withholding. Options may not be obvious. Once they have a sense of what they can control, they may be more likely to participate. It is important to allow them to suppress or limit personal information in directories and bios.
Explicit support for multimembership	• "My communities" page • Single identity (login, profile) across communities • Aggregators (RSS, tags, feeds) • Lists of communities on personal pages	This is perhaps one of the greatest opportunities for technology innovation.

Chapter 6. Community orientations: activities and tools

levels of commitment, they take on different roles, and they use tools differently. The community and its learning mean different things in their lives. They develop distinct identities as members and express their relationship to the community in their own ways.

Communities vary in their degree of orientation to individual participation. They make more or less effort to accommodate individual differences, recognize multimembership, or take advantage of their diversity. In bringing people together, some communities offer only one way to interact, regardless of individual preferences, in order to create a shared history of interactions. Others offer a wider range of interaction possibilities and styles, accommodating individual differences in participation but loosening the bonds created by common interaction experiences. Global communities need to accommodate diverse time zones, languages, and cultures.

This orientation to individual participation has both private and communal dimensions. It enables members to take active control of their participation, and it makes individual differences part of the life of the community. The main variants of this orientation include:

- *Varying and selective participation*: Communities accommodate various forms of participation, ranging from just staying lightly in touch, to choosing a few areas of personal interest, to participating actively overall, to taking a leadership role.

- *Personalization*: Members can individualize their experience of the community to serve their personal needs and circumstances and control access to their information.

- *Individual development*: The community helps individuals develop their own learning trajectories, through guidance, mentorship, and individualized resources.

- *Multimembership*: Belonging to multiple communities and managing participation across these contexts is a fact and a challenge that can remain private or be expressed outwardly in the way a community organizes participation.

Signs of life:

Members develop their own style of participation and are aware that other people develop other styles. They feel they can have a meaningful connection to the community what-ever their individual form of participation, and the community welcomes, supports, and thrives on this diversity.

Key success factors:

- Diversity is explicitly valued.

- Different levels and modes of participation are supported and facilitated.

- Practices and tools are used to bridge between interaction modes (audio, text, video, synchronous, asynchronous, face-to-face, online).

- Preferences, availability, and multimembership can be communicated.

- Customization options are obvious and understood.

- Members can manage their interactions across different tools and multiple communities.

Questions to consider:

- To what extent does the community's success depend on uniform participation expectations, such as logging on to an online space daily or weekly, regular meetings or interactions, and scheduled events?

- What is the degree of diversity among members in terms of level of proficiency in the community's core practice, as well as members' literacy, learning styles, language, culture, and access to and familiarity with technology? Do members have strong and different preferences about interaction modes?

- How much ownership do members take or want to take of their own learning and development compared to how much they expect this to be defined by the community as a whole?

- How many communities do members belong to simultaneously? Are they all within one organization and therefore use the same set of tools?

Technology implications:

When technology becomes the members' main window into their communities, their participation can be a highly individual experience. This participation may consist of a series of visits to a website or to web conferences. Or it could be participation in a variety of online events, conversations, and meetings. Communities need a technology infrastructure that can translate this succession of points of contact into a meaningful experience of participation over time. This is especially important if various modes of interaction are supported with different technologies. Bridging them is critical to keeping the community together while enabling various modes of participation—for instance, offering members the option of having information pushed to them via newsletters and email alerts or allowing them to selectively organize how they access content.

When the intensity of participation varies a great deal among members, those who participate infrequently or superficially can be overwhelmed by new material and new topics. In this case, it can be important to have individualized guideposts such as member-specific new flags or pointers that reflect the member's interests.

Multimembership and individual expressions of identity are taking on increasing importance as technologies multiply the possibilities for simultaneous participation in communities. Members need configuration options to manage their participation and attention across more than one community with a single set of tools. Many potential members balk at the idea of having to learn a new set of tools or to remember another user id/password for each community.

Orientation 8: Community cultivation

While many communities are happy with loose self-organization and unplanned evolution, others thrive on attention to community cultivation. They have a need to reflect on the effectiveness and health of the community to make things better, joined with a willingness to work on it. Sometimes regular members are more interested in the domain, and attention to the work of cultivation is the province of a smaller core group or one person. Such leaders facilitate conversations, convene meetings, organize activities, collect, edit, or produce resources, connect members, keep a pulse on the health of the community, and encourage it along a developmental path. Whether these people are volunteers or paid members, the success of the community comes to depend on the high level of ongoing attention that these leaders pay to process and content. The main variants of this orientation to cultivation include:

- *Democratic governance*: Some communities create governance structures and processes that enable the membership to have a voice in running the community, engaging in self-design.

- *Strong core group*: A distinct group of members habitually take a nurturing role with their community.

- *Internal coordination*: A member or a small team explicitly takes on or is assigned the responsibility of cultivating the community.

- *External facilitation*: Someone who is not a member is recruited to provide process support to the community. Such a person may not be knowledgeable or even particularly interested in the domain, but is assigned this role because of expertise in community cultivation.

Signs of life:

The community's activities are well planned, its reference materials are well produced and well organized, and members find that someone is always very responsive to their requests, contributions, and changing needs.

Key success factors:

- Efforts made to support the community by members are appreciated by other members.

- Enough time available to engage in cultivation.

- The personality, skills, leadership, and reputation of those who take on cultivation roles in the community.

- Succession planning for transitions.

Questions to consider:

- What information do community cultivators need about the activities, workings, and health of the community? For example, is there a need to track participation, downloads, and usefulness of content?

- What actions should cultivators be able to take with respect to technology? Who should be given the privilege to control other people's participation? How much time and willingness do community cultivators have to devote to learning how to use sophisticated cultivation tools?

- What is the community culture around feedback tools? What are the effects of making that participation visible? Is there a risk of people "gaming" the system to affect outcomes of things such as rating systems or polls?

Careful cultivation as key success factor for on-line community

The founders of the community CompanyCommand had not planned to become community cultivators. They were company commanders in the U.S. Army and had benefited greatly from informal conversations they were having about the challenges of leadership. They had written a book to capture their learning and had launched a website to discuss the book with colleagues. The site became the focus for a rapidly growing community of officers who were in the two-year "company command" phase of their careers, plus a good number of others who were either preparing or had moved on. The site now reaches several thousands.

The nature of the domain—perhaps the most challenging phase of an officer's career—and the camaraderie inherent in the community naturally make for a strong sense of identification among the members. They need help with their practice and are more than willing to give help. Yet it is hard to overstate the amount of cultivation work and skill that goes into making sure that the site provides a valuable learning resource for practice; it is constantly being renewed by the community.

The four founders and other volunteers form a dedicated cultivation team. For instance, they read and clean up every contributed document before it is posted. They always thank contributors individually. They interview leaders and practitioners and edit these interviews into downloadable videos. They offer a quiz of the month that challenges members with a practical problem to address and discuss. They publish a monthly column in a publication with the best of the community.

They have cultivated a core group of about 150 volunteer topic leaders who each facilitate a topic area. They have given topic leaders business cards to recognize their role. They also organize a yearly face-to-face gathering for them, for coordination, learning, companionship, and recognition. They make sure they are grooming a new generation of community leaders.

Through their constant cultivation work, the community leaders make the experience of contributing and participating very personal, useful, up-to-date, and based in practice. It is for practitioners by practitioners.

The CompanyCommand site, which was public for several years, is now behind a firewall and no longer publicly viewable. But the founders have described their cultivation effort in a book with lessons and advice that are applicable far beyond the military context.[5]

---

5. Nancy M. Dixon, Nate Allen, Tony Burgess, Pete Kilner, Steve Schweitzer, Company Command: Unleashing the power of the army profession. (West Point, New York: Center for the Advancement of Leader Development & Organizational Learning, 2005)

Table 6.8 - Activities and tools for community cultivation		
Activities	**Tools**	**Practice Notes**
Announcements, stories, pointers, and other information sent to members directly	• Email • Newsletter • Community blog (internal) • Calendar	More can be less. Consider members' attention spans. If you push too much information, people may start to ignore it.
Getting community input and feedback	• Polling tools • Brainstorming tools • Email • SMS	Consider making the tools available for members to create their own queries and questions.
Backchannel communication, offline conflict resolution, and private encouragement	• Membership contact information • Phone • IM • Email • Chat (during meetings, for example) • SMS • Microblogging	Personal touch is still the best way to focus attention, connection, and participation. Most of us respond positively to personal contact. This requires community leaders' time, more than any specific technology.
Reflecting on community participation and health	• Participation statistics • Alerts noting lengthy member absences • Community health charts (indicators of level of participation, quality of conversations) • Social network analysis • Logs of technology use, such as when people have logged in, how long they stayed, or how much they have read • Lists of who has read or down-loaded something	In general, we advocate making this kind of information available to community members because it supports a reflective practice. Again, one has to be sensitive to the community's culture and the degree to which information about individual participation helps the community or becomes a distraction.
"Housekeeping" inter-actions	• Ability to move contributions from one place to another to keep an online space organized (for example, moving a post to a different conversation or a document to a different folder) • Tracking an individual's contributions across contexts • Conversation analysis tools (for example, contributions that open or close threads) • Access lists	A facilitator of an online discussion can help keep focus and cohesion by moving misplaced posts, noticing who is contributing in different areas and who is quiet and might be encouraged. Community leaders need to be very careful about the use of such tools because they reflect authority and power. Frequent communication is recommended, such as an email to the author of a post that needs to be moved.
Rewarding behavior valued by the community	• Top contributors or "member of the month" • Quantified reward system (for example, points for certain behaviors)	Rewards can be a double-edged sword. They can encourage participation or feel manipulative by encouraging only specific behaviors. Consider the community dynamics carefully before implementing rewards.

Technology implications:

General communication tools such as phone, email, and instant messaging are still the basics of community cultivation. A lot of community cultivation is simply about keeping in touch with members through backchannel communication, where people communicate privately amongst themselves. A phone call can be effective, as can a quick IM when someone is online to say "hello" or encourage participation, particularly in smaller communities. Broadcasting tools help keep people informed about community activities. With a palette of available tools, cultivators can customize communication to the person or the context. Intensive cultivation also calls for more specialized tools to poll members, brainstorm ideas, or manage conversations, documents, and archives. Finally, some tools can help cultivators "see" the community by tracking participation statistics including logins, pages read, contributions posted, and downloads.

These tools can help identify current topics, track individuals' contributions, and be used to chart who is engaged and who may need encouragement or be "invited back." Again, the availability and use of these community visualization or evaluation tools raise issues of privacy and availability of information. Who can see the information? How it is used? Does it help the community or does it create unwelcome distinctions among members, such as between those who are recognized as active contributors (so are valued more) and those who only read or participate in less noticeable ways?

Many available tools can generate large amounts of log data; the challenge is to integrate information from different tools and to reduce it to something simple and easy to act on. People who have an explicit role in cultivating a community are more likely to take the time to learn how to use the tools. However, community-cultivating tools can be useful for any member who cares about a community.

Orientation 9: Serving a context

All communities of practice are oriented to their members' learning experience. They always exist in a context that, to some extent, influences how this learning takes place. But in some cases, serving a specific context becomes central to the community's identity and the ways it operates.

Some communities are not especially oriented to serving a context: the members mostly seek intimacy and privacy and the ability to interact and share materials far from the public gaze. Their agenda is an exclusive focus on the learning of members. But many communities of practice are defined by their orientation to serving a context beyond the learning of members. They may live inside an organization, whose charter their practice needs to serve. They may have a mission to provide learning resources to the world or to recruit members widely. Or they may seek interactions with other communities whose domain complements their own. This outward-facing focus can become a key driver of the community's evolution, a selection criterion for members, and the inspiration for partici-pation. The main variants of this orientation to context include:

- *Organization as context*: Communities living within an organization usually feel a responsibility to develop capabilities that serve the charter of their host organization. Organizational membership may be a condition for community membership and a key to trust. Such communities may also need to use that organization's resources and infrastructures and worry about interoperability, integration with the organization's operations, and interaction with its power structure. They may be focused on shaping organizational strategy or practice.

- *Cross-organizational context*: Some communities find value in creating connections among practitioners across organizations, without the necessity of forging more formal relationships among these organizations. This context creates its own set of relationships to these organizations' charters, resources, and power structures, as well as issues of communication across firewalls and platforms.

- *Constellation of related communities*: Some communities need to constantly interact with other communities to form broader constellations and networks. They need to negotiate related domains, seek interactions at their boundaries, encourage multimembership, and coordinate their learning.

- *Public mission*: When a community is built on a mission to serve the broader public, it needs to interact with entities and individuals outside the membership. This often entails creating specific resources and activities to make the learning of the community intelligible and accessible to non-members.

Orientation to an organizational context embedded in technology

In a large company like IBM, communities of practice easily number in the hundreds and are closely integrated in the organization: some are specifically sponsored to address a strategic domain; others are simply recognized as part of the way things are done. But all are expected by members and management alike to contribute to the business through exchanges, tools, ideas, and strategic thinking. This tight integration is reflected in technology. The company has been designing a program called CommunityMap, which serves as a kind of online registry for communities. The idea is to give each community a page in the registry with a description of the domain, the membership, and the leader(s) to contact. If someone wants to start a new community, the first step is to make sure that a community doesn't already exist for the same or a closely related domain. But the integration goes beyond a mere registry. The program also supports the integration of membership management with other systems. For instance, each community is assigned an IM channel that all members are automatically registered to use. Community membership is indicated on each member's employee page in the HR database. When employees leave the company, all their community memberships are automatically terminated by default.

Signs of life:

Community members are fully engaged in the mission defined by their context. Reciprocally, recognition and resources come from people outside the community.

Table 6.9 - Activities and tools for serving a context		
Activities	**Tools**	**Practice Notes**
Creating a public face for the community	• Public, searchable web pages • Community blogs (external) • "Friends of the community" email lists • Public newsletters	Even a community mostly oriented to the learning of its members can benefit from having a "public face," if only to make others aware of its existence.
Inviting the public in and recruiting members	• Public areas • Guest accounts • Self-registration	Technology can mark a clear entryway as well as provide tools to support those who are welcoming new members. However, technology is not the only thing to consider—a personal invitation is often the most effective.
Offering community content out to the world	• Web support for publication streams • Search tools • Meta-data • Tagging • RSS feeds	Making community content useful to outsiders may require repurposing internal content, with a dedicated editorial staff. Any content offered to the public in this day and age should have an RSS feed at a minimum.
Knowledge transactions for non-members, help desk	• Question-answer systems • FAQs area • Phone • Email	This is a version of Orientation 5: Access to expertise, but offering this service to non-members usually requires a more organized help desk to handle requests and protect the time of members.
Constellations of related communities	• Shared community portal • Community mapping tools	When existing resources can serve more than one community, there is more reason to harmonize at least some of the technology used across communities.
Backend compatibility with organizational infrastructures	• Single login systems (LDAP/Active Directory) • Standards (databases, XML, .NET) • Look and feel of the user interface • API/web service • MAPI and directory structures	When existing IT systems provide an important part of a community's context, reaching out to the IT people makes a big difference. Tech stewards need to partner with people who have greater technical depth and authority over security and other infrastructure.
Security	• Password protection • Access management • Firewalls	Security concerns may not really be the business of community leaders, but often the activities of communities can raise unique concerns from a security perspective.

Key success factors:

• Clarity on the community context and its implications.

• Channels for negotiating the relationship of the community to its context, such as organizational sponsorship or good connections among community leaders.

• Recognized and supported boundary roles that serve the orientation to context.

• Tools that enable outsiders to interact with the community in ways that reflect both their needs and the community's desire for openness.

• Ease of granting controlled or open access.

Questions to consider:

• What goals, agenda, or mission is the community serving? What aspects of the community does this determine (for example, learning goals, membership, or assessment)?

- To what extent does the community have to keep track of its activities and its learning to justify its existence to outside constituencies?

- How important is it for the community's technology infrastructure to be integrated within broader information systems?

- Is the community open or closed? Is there a specific membership procedure or set of requirements, or can anyone join? How does the community attract new members? Is it necessary to have a strong "external face" to create that invitation?

- How important is it to make the community visible and/or accessible to non-members? What would these people need? What other communities is the community connected to or "related" to, and how do they currently interact?

- How do members integrate their activities in the community with their other activities, such as their jobs in their organizations?

Technology implications:

The degree to which a community's context is central to its identity creates specific technology-related challenges. Within an organization, it often requires compatibility with the existing infrastructure. Single login and "closeness" to the tools members use in their daily occupations can also facilitate participation.

A broader orientation to serving a context calls for specific tools that provide for an outward face to the community and affords choice in how boundaries are defined and maintained, as well as transactions across community boundaries. This orientation can require either open or closed systems. Those who seek intimacy and privacy need tools that create strong boundaries, while those with an open face to the public need the ability to be visible and to interact and share materials outside. For example, a public context suggests avoiding passwords and other barriers that prevent public search engines from indexing content. An organizational context may require passwords to protect intellectual property but provide access to anyone with an organizational password. Many communities have both closed areas for their own internal work and open areas for their interactions with the outside.

Using orientations to think about technology needs

The framework of community orientations is useful for thinking about the technology needs of a community because it places technology in the context of the community's patterns of activities. Depending on how technology stewarding is organized in a community, these orientations can be used in several different ways. In some communities, such as those where technology is a common interest, the entire community gets involved in discussing the orientations and considering which ones are relevant. In other cases, a small group will think about the orientations and the questions they raise, and then engage the whole community in considering the results. For such an evaluation, the variants, signs of life, and questions to consider associated with each orientation provide a framework with implications for technology choice.

When a community is just forming, its profile of orientations may not yet be apparent, so a tech steward has no history to go on and can't really say what orientations are most descriptive. In this case, orientations can be used to trigger the imagination of potential members in projecting what their community will need. For an existing community, the use of the framework will depend on whether the community is happy to maintain its existing style; the framework provides an analytical tool to assess how well the community is being served in its current form. If a community is seeking change, then the framework of orientations can provide a language to imagine the future, discuss newly evolving needs, and put technology to work in the service of the community's intended evolution.

The main idea is to create an actual or intended community profile in terms of orientations and their variants. As you explore each orientation and variants listed above with each orientation, think about how closely they apply to your community, using a scale of 1 to 5, with 5 being very important. If your time, attention, and budget as a tech steward are limited, focus first on the orientations you rate 4 or 5. That way, the orientations profile provides a useful reference point for the task of prioritizing, selecting, configuring, and even supporting tools.

In Chapters 4 and 5, we looked at technology from the perspective of the full landscape of tools and how features, tools, platforms and configurations come together to support a community. We asked, "What can a specific technology do for our communities?" In Chapter 6, we looked from the other direction, using orientations to help us look at a community's technology needs through the perspective of the real activities of the community. Together, these three chapters give us the language and framework to see technology from a community perspective and community from a technology perspective. And by looking through these perspectives, we can both imagine and support the useful application of technology for our communities. With a language to talk about digital habitats, we are ready to dive into the practice of technology stewardship.

Part III: Practice

Assessing your community context

In this chapter we describe many contextual factors involved in making the pragmatic technology choices that shape a community's digital habitat. These factors are important because neither communities nor technologies exist in isolation.

Your community's state of readiness for technology change

The assumption that a community needs to change or use more or different tools should not be taken for granted. Several factors can affect just how much change in technology is needed, and whether change will be tolerated, much less welcomed.

- **Stage of community development.** A community's technology needs vary over its lifecycle. Is your community new, rapidly changing, or settled? Mature communities may have more of a sense of who they are and what they want. There may be specific feature and configuration requirements that inform technology acquisition. Your community simply may be restless and ready for change, or settled and resistant to change. Restlessness will encourage a tech steward to look for new technologies. Being settled but with new needs calls for the selection of technologies that are only incrementally different from the current community configuration.

New communities are likely to be uncertain or not yet clear on their orientations. They need to imagine what they might become and leave their options open. Community members may bring in a diverse set of favorite tools with them, making the early focus on coordinating the use of those disparate tools rather than moving to a completely new platform. In a technology-savvy community, members may have strong ideas about adopting tools they prefer. As a community coalesces and matures, it becomes easier to involve a broader group of members or the whole community in technology decisions.

- **Community diversity and complexity.** Communities and their needs vary in complexity, which affects technology choices. How many orientations are important to your community? How diverse is your membership? How geographically spread out is it? Too many tools or features may dampen, rather than enable, activities. Complex orientation sets may suggest the need for more tools. Many platforms can accommodate multiple languages in their interface (for example, menus or static content that has been manually translated), though interactions in more than one language are still a challenge.

- **Community experience with the use of technology.** Technology use depends on a combination of access, skills, and preferences. How experienced and skillful are members with technologies? In communities with a solid base of technical skills, it is easier to experiment with new tools so people can learn, evaluate, and then adopt or discard them. In other communities, members may be uncomfortable moving beyond email.

- **Community attitudes about technology.** What level of innovation energizes, rather than discourages, your community? What past history of technology change in the community might inform new changes? When you ask a question about technology in the community, do members reply with interest or ignore it? Members passionate about and invested in their domain will be more willing to devote the time necessary to learn to use new tools that support their collective work. Without a compelling reason to justify the investment of time, new tools and the technology skills they require are likely to be considered a distraction.

Your community's relationship with its environment

A community's accountability has direct technology implications, as does its desire to be open to the world or to be private. Some communities have to consider both the external and internal effects of their technology selection. They may want to share their learning with the whole world. Others are only responsible to each other and only want to share internally.

Here are some factors to consider about a community's environment.

- **Organizational relationships.** How beholden is your community to an organization? Is it completely "contained" within an organization? What role does the organization have in technology decisions? What technological constraints and opportunities does this relationship create? How much of your community needs to be interconnected to the day-to-day activities of other organizations (customers, suppliers, partners, funders)? How much privacy from the larger organization does the community need?

- **Relationship to an IT department.** Communities may have specific relationships to IT providers or departments, either because of a contract for services or as part of provisioning offered by a sponsor or host organization. These relationships often define a set of resources and constraints to a community's technology decisions. Including the decision makers and technologists before finalizing technology decisions is important. They can be helpful selection partners, and compatibility with their systems will make the implementation and provisioning of technology easier in the long run. However, they can also be so focused on technology that they undervalue community requirements. Representing the voice of communities sometimes involves serious negotiations.

- **Need for connection to the outside.** Different communities choose different levels of connection to the outside world. Some need to be private and secure; for others, openness is a key to the life of the community. Communities that serve the world need tools that allow the world to see and interact with them. Communities with issues of intellectual property or security need strong boundaries. Communities that interact with other specific constituencies need tools that work across boundaries.

- **Multimembership.** Your community members probably belong to multiple communities—some related and some unrelated to yours. As individuals manage their membership in these communities, they also have to learn to use a potentially wide array of tools. How can your configuration make it easier for members to deal with this challenge? Do you offer tools and features such as RSS feeds that allow members to harvest materials from your community for reading with materials from their other communities? How much weight does interoperability have in your technology decision making?

- **Security.** How should your community address security? Two aspects are important: securing member and community information, and protecting the community from spamming or other external incursions. A public-facing community may have a strong interest in reducing the impact of spam. A more inward-facing community may feel a strong need for privacy.

  Security features usually are an integral part of a platform offering. In looser configurations, tech stewards may have to devise a security approach—in terms of both technology and practice—that works across and between tools, and understand the security characteristics of each tool in the community's configuration. For example, Skype conference calls (VoIP calls) are secure because they are encrypted, but a recording of a conversation can easily be copied and forwarded inappropriately.

Time and sequence issues

Choosing a technology creates a moment in time when a community moves forward most visibly. Making a choice on behalf of a community is different from scanning, where you can shop around and consider many alternatives but not feel the pressure to act. It's also different from your work to support the community's use of technology, which is also ongoing. With this spotlight on technology stewarding, it is important to consider how to plan and coordinate technology changes.

- **Community schedules.** Depending on the complexity or magnitude of the technology decision, it is useful to create a schedule for technology change. This might include recapping the scanning process or spelling out the details about when and how selection and implementation will occur. A schedule helps reveal how interconnected many of the decisions are, and give you clarity about where you do or do not want community input. For example, announcing a cutoff to your search can prevent endless deliberations and give everyone a sense of clarity about when the discussion is over. Remember, technology change is not the domain of the community and too much focus on it can be disruptive. Not enough focus can leave people feeling disenfranchised, so coordinate your technology schedule within the larger perspective of the community's life.

- **External schedules.** Consider how your community's schedule interconnects with the schedules and rhythms of sponsoring organizations and other outside factors. A purchase decision can have budget implications that must fit with a budget cycle. A major changeover of technology may not be advisable on a Friday when tech support goes home at 5 PM. A push to get people to try a new tool may be disrupted by a holiday.

- **Not your last chance.** Although your current decision may loom as the largest and most consequential decision for your community's technology, remember that there will be other big decisions that will need to be made in the future. Therefore deferring some issues until some future time is completely legitimate and appropriate.

Your community's budget and resource considerations

Investments in technology require tech stewards to make realistic assessments about the availability of financial and other resources. A lack of money or time can slow the process but it can also stimulate the community to find free resources. Financial sponsorship may stimulate ambitious technology plans but force communities to justify the investment. These resource considerations must be balanced with the goals and aspirations of the community.

- **Scope and budget.** Before you make a selection, decide on the scope of the project. How big or small do you want to start? Is money available for purchasing tools, custom development, and technology stewarding, or to pay for hosted applications? Having no money does not mean having no options, but it suggests adopting certain strategies that minimize costs. What are your short- and longer-term goals? What is critical now and what can wait? Having this information on hand helps you choose incremental technology adoption strategies to ensure you start at the right level and grow from there. It also helps you look for pricing models in the market that match your budget.

- **Contributors, decision makers and stakeholders.** Knowing who might have a say in technology investments—such as sponsors, organizational hosts, donors, or even individual community members—and how you want to work with them, can help unlock those resources. You may need to gather data to justify your expenditures. The types of evaluation suggested in this book help you to clarify why you have selected certain technologies and what their value is to the community.

- **Technical resources and expertise.** How much technical know-how do you yourself have? How much external expertise (for example, from outside tech stewards, or other organizational resources) is available? If technical expertise is scarce, choose technology options that require little programming, such as hosted applications, or select simple tools that members already use. If the community has complex requirements, then you need to find the expertise. If you have expertise but no need for complexity, don't get carried away building things that are not necessary or appreciated.

- **The tech steward's time.** The resources noted above are all important, but so are you, the tech steward. How much time do you have to shop for and implement technology? Use your own availability as a guide to the level of technology investment and complexity you choose.

Technology infrastructure considerations

Communities are influenced by the technology infrastructure that's available to the community as a whole and to members individually. Communities situated in organizations with IT departments are faced with requirements or opportunities that may not exist for others.

- **Online access and individual technical setting.** Many of the tools communities use assume good Internet access. Some even assume that users have broadband Internet capacity that's always on. How do your community members access bandwidth—from the office, from home, or from a mobile device? How much online time do they have? Are members using PCs or Macs or both? Does the software run on just one or on both platforms? How powerful must members' computers be for them to have a satisfactory experience? How important (or feasible) is it to accommodate varying conditions within your community? For example, if members are scattered across the world in diverse socio-economic settings, they may have very different technical infrastructures. Even members from within a single organization may have different conditions (for example, some may participate from home).

- **Hosting and vendor relationships.** Communities have to make choices about the software they use and how it is hosted. Some may purchase software from a vendor while others use hosting services from ASPs. Both these choices and the relationships with the vendors are worth consideration.

  - **Application service providers.** Application service providers, or ASPs, offer customers both the software and the server hosting. If you don't have a server or the energy and expertise to maintain one, hosted applications are very useful. Today, externally hosted applications can be just as secure as applications hosted on your system. ASP options range from free (mostly ad-supported or with limited functionality) to full service, with varying price tags. Some vendors offer extensive customization of their product; others offer none. The advantage of ASPs is the ease of implementation. The downside is that your data is locked in someone else's system, which may not be interoperable with other tools your community or organization uses. Ask the ASP about other communities that have used its service. Find out about the ASP's uptime and reliability.

  - **Hosting your own.** You can host applications on your own server but someone in the community needs to have the knowledge to install and support them.

  - **Working with software vendors.** Vendors can be useful partners for communities, particularly if they are interested in developing their product based on feedback from you and other customers. They can also be persistent sales people only interested in the sale. Consider what sort of relationship you would like with a vendor and how ready they are to work with you. Find out how long they have been in business and how long the product has been available. Is support provided? Does a support community exist? What other communities have purchased their software? Ask if you can speak to some of those customers.

- **Programming.** Questions to ask a software vendor related to programming are: Will the operation of a platform require ongoing programming and if so, what type—actual code or HTML templates for appearance? How important is configurability and to what extent is the platform under consideration configurable? Can it be tweaked and configured by a community member or by the tech steward, or does modification require the help of a skilled technologist?

- **Internal IT resources and system requirements.** If your community exists within or is sponsored by an organization, your technology choices probably will need to be assessed and accepted by your IT department. Talk with that team before you settle on your acquisition strategy. If you are constrained to use only particular technology(ies), there is no need to scan for others. Instead, you should focus on making sure that your community gets the most out of the technologies your IT department accepts.

If you need a special piece of software, prepare a strong case for why it is necessary. For example, some companies have concerns about downloading and installing synchronous technologies, such as IM or Skype, because of security and bandwidth issues. Community members may consider it a given that these tools allow easy voice communication and save on phone costs, without understanding IT's concerns. Tech stewards can open the dialog around these issues. Explore questions of back-end software compatibility with the organization's other applications, including firewalls. Will your desired software work with or against existing applications? Does the organization have standards around supported hardware, operating systems, databases, or other software characteristics? These standards may restrict your selection, although the trend toward ASPs and the increasing use of open standards like Java and XML decreases the prominence of these types of problems.

- **Client programs.** Some systems (such as eRoom or Groove) require the download and installation of software on a user's machine. Most programs that allow sharing of applications require at least a Java applet. Your organization's IT department may need to do the installation. Browser-based or thin-client applications are increasingly common and do not require local technical support, but some organizations ban downloading and installing these.

- **Single sign-on.** Consider the implications of members having to sign on to multiple systems to participate in the community. Many communities find multiple sign-ons to be an irritant or a barrier to participation. For technology-savvy communities, the benefit of diverse tools may be so important that multiple sign-ons are not a problem. If you use a separate system, determine if you can integrate it with your members' day-to-day applications. If you choose single password, make sure the IT department approves because typically the IT department controls the sign-on protocol. For example, one organization used the open-source WordPress[1] blogging tool and connected it to their enterprise user registration database to provide access to both tools with a single sign-on.

A note on platform pricing

Pricing structures are a technology infrastructure concern that can subtly affect how communities develop and how individuals participate in their communities.[2] For example, a platform price that is structured "per community" may inhibit the quick launch of

Table 7.1. Types of pricing structures for community software			
Pricing Basis	**Advantages**	**Disadvantages**	**Examples**
Per community Incremental cost for each additional community, with or without limits on membership and activity	• Good for stanalone communities • Large and open membership • Predictable cost • May include design, support, and facilitation	• Costly if you start having many small communities • Need to have a clear sense up front of the value a given community can deliver	• Communispace • Tomoye Ecco • CommunityZero (per community and by volume)
Per seat Incremental cost for each person, no matter how many communities they participate in	• Good for systems of communities, especially inside an organization • Enables communities to be started easily "on demand" • Allows anyone to participate in the communities and at the level they choose	• Limits membership and peripheral participation to a "paying" group • Expensive if some people do not use the system much or at all	• SiteScape • Groove (also offers volume pricing) • Socialtext • WebBoard
Per volume of activity Incremental cost for additional use; the system calculates actual usage, usually in terms of page views or transfer of information	• Enables communities to start without having to demonstrate value up front • Peripheral participants can be included without "taking up" a seat • Good for interorganization communities	• Hard to predict cost • Requires communities to have a license that allows you to grow as you go, because estimating volume from the beginning can be difficult	• Web Crossing • BaseCamp (per project) • Many ASPs
Per platform Outright purchase of the software, site license, or, for an ASP, "unlimited usage" contract	Full freedom to: • Enable creation of communities "on demand" • Support various forms and levels of participation • Free for open source	• Potentially high start-up cost for enterprise-level tools • Inexpensive off-the-shelf systems require hosting capabilities	• SiteScape DiscusWare Pro • SharePoint Portal Server • Drupal (free) • Wikispaces

---

1. WordPress http://wordpress.com
2. Pricing structure information is subject to change, but what we show here is accurate as of the time of publication.

subcommunities or small communities that are still tentative about their viability. If the price is "per member," how do you take into consideration peripheral members who may be discouraged from participating? What if the pricing includes hosting? Does that free your community from having to shoulder internal IT costs or worries?

Four types of pricing structures are outlined in Table 7.1, along with their advantages and disadvantages. Most vendors have adopted hybrid structures. iCohere, for example, charges by community, with a seat limitation. CommunityZero charges by community with a volume (bandwidth and storage) limitation. Web Crossing[3] charges by usage up to a threshold volume of activity after which the license is unlimited. SiteScape offers a variety of options that make sense for an organization with multiple communities. Many of the vendors have additional pricing options for support, hosting and other services.

It is important to take a community's evolution into account. For example, a solution that's great for a small, tightly knit community of twenty people (based on a cost per page view per month) may become prohibitive when the community grows and costs rise dramatically with increased page views. Vendors may be open to negotiating different options. For example, those who offer one pricing structure as a default will sometimes negotiate another. If a vendor does not offer the pricing model or the package the community wants, it is worth asking the vendor for that option.

Because of the complexity of a community's digital habitat, a tech steward may need to pay attention to a daunting number of things. Try using your community's needs in order to focus your attention on the most important and most practical issues. Track your work with a journal or blog of issues. A record of your experience can prompt insights from others who are involved in stewarding technology for your community and who might be able to help you if they have enough context. The bottom line is that you should try to to stay aware of the context of your community as it evolves and understand how that context affects your technology stewarding. Your community's context will influence decisions about technology acquisition strategies, as discussed next.

---

3. Web Crossing, www.webcrossing.com

Technology acquisition strategies

Which technology acquisition strategy is appropriate depends on the circumstances of an individual community. This chapter explores strategies that change a community's digital habitat. Our exploration includes budget, comfort with technology, and availability of other resources.

Technology stewarding is most visible when it's time to choose something. All of a sudden, people pay attention to you because decisions are being made, resources are being allocated, and something is about to change in the community. As a tech steward you need a strategy for this very visible stage that takes into consideration all the things you know about your community and about the options and choices in the marketplace.

We suggest seven general acquisition strategies as shown in Figure 8.1. Each strategy has its pros and cons, and each implies a relationship with certain actors—something that matters when you are in the position of asking for help or for permission.

Figure 8.1. Seven acquisition strategies

Based on your orientations and context (discussed in Chapters 6 and 7), you might focus on just one of these strategies, or you might combine them in various ways. In describing these strategies, we have numbered them for convenience, but neither the numbers nor the order is significant.

One way to look at choosing a strategy is to address each of your major community orientations to make sure that a tool or platform you are thinking about meets functional needs. Then consider your community's context. Finally, consider the acquisition strategies.

Consider the example in Figure 8.2. Imagine that you identify four important orientations in your community: meetings, content, open-ended informal conversations between meetings, and relationship. You know that you prefer open-source software, you need to respond to the corporate IT department's requirements, and your members are pretty technology-savvy. You assess what you have and what you need, and then go see what the market has to offer. You decide to build on the corporate intranet and add some addi-

Chapter 8. Technology acquisition strategies

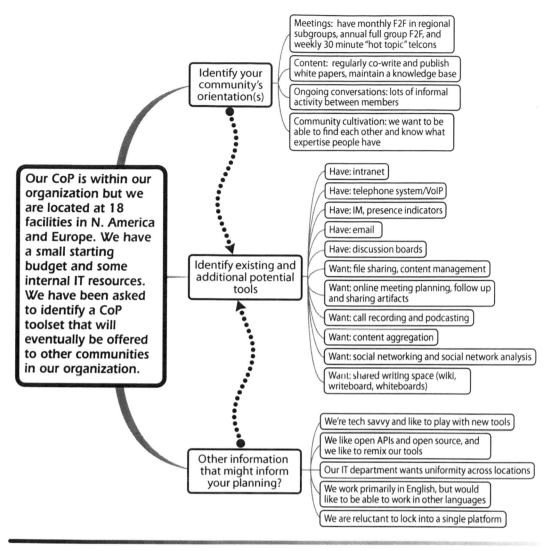

Figure 8.2. How community orientations and context can interact with acquisition strategy

tional functionality—some integrated into the corporate IT structure and some hosted externally by a provider.

Strategy 1. Use what you have

Communities of practice existed long before the current crop of technologies became a factor in community life. Your organization or members of your community may already have access to all you need to get your community going and are familiar and adept with these existing tools. An opportunistic "borrow, adopt, adapt, and cobble together" strategy may require that you take a second look at what you have and use your tools in a new way. Simplicity is a real plus, especially at the beginning when commitment to the community can be tentative. This approach may not be very exciting for technology-savvy members, but it is not to be dismissed out of hand. It builds on relationships with anyone who can offer a tool. Figure 8.3 offers a snapshot of what this might look like in a community.

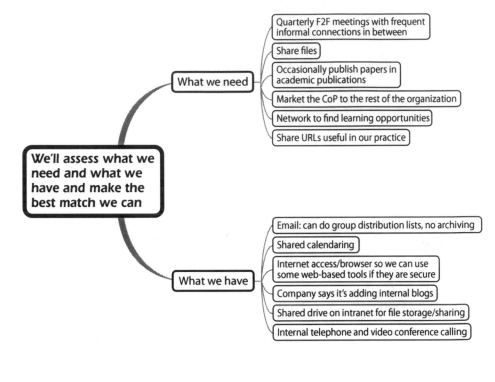

Figure 8.3. Use what you have

- **Pros:** This strategy requires no budget and little or no negotiations with an IT department. Using this strategy means avoiding the learning curve that a community-specific platform would entail for members because the tools are already in use. Outside an organization, it creates commitment when members contribute facilities they have access to.

- **Cons:** Across organizations, firewalls can be a problem. Downloading external software may be forbidden. The availability of a resource may depend on the continuing membership of the donor. In an organization, this strategy does not imply a commitment to supporting a community.

- **Tips:**
  - If your community already exists, find out what is currently working and think about why it is working.

　　　　　　　　　　　　Chapter 8. Technology acquisition strategies

- Explore new or more effective ways to use existing tools in service of the community.

- Assess what's available from member and sponsoring organizations.

- When a tool from a single source may be transitory or uncertain, consider the benefits of having back-up tools that essentially do the same thing.

- If you discover you have access to enterprise-level resources, check out strategy 3.

- **Examples:** The basic communication technologies that most organizations use can be enough for some communities. Email systems usually have facilities for creating simple distribution lists. Most organizations have some kind of file repository system. Facilities for telephone conferences are commonplace.

Strategy 2. Go for the free stuff

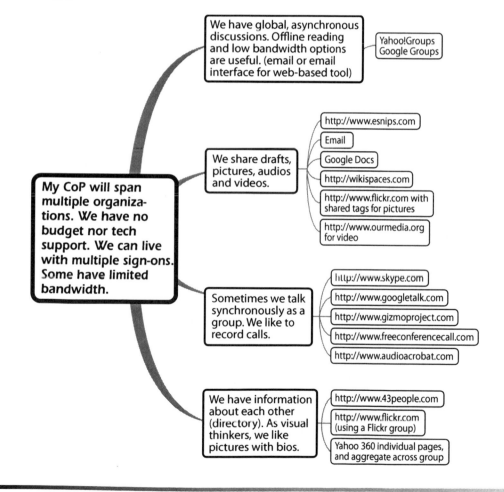

Figure 8.4. Go for the free stuff

A growing selection of free community tools and technology is available for communities with limited financial or technology resources and the need to get something going quickly. Because they have no dedicated IT resources, communities that span organizations or exist fully outside them may find this approach particularly attractive.

This strategy implies a dependency on the providers of free tools but most likely very little relationship with them or influence on the development of the tools.

- **Pros:** Some freely available systems are very easy to set up and use, and you can't beat the price. This strategy offers low-risk experimentation. The tools and features available often rival those found on pricey platforms. Technology-savvy members may find it liberating to add and experiment.

- **Cons:** Free systems are often supported by advertising or by tempting you to upgrade to a more feature-rich version of the tool. Cobbling together several free systems may create integration issues and convey a weak sense of identity to members. There may be questions about the security and control of community information when you do not control the platform. You may have to jump between varieties of free tools to get everything you want. Less technology-savvy members may not have the skills or inclination to test and jump between tools.

- **Tips:**

   ○ Back up copies of your data (particularly your member data) somewhere external to the free web-based platforms in case they go down or go out of business.

   ○ Consider how important archives and control are for your community. If control is essential, free tools may not be your best option.

   ○ Consider blending free tools with other tools you already have or can acquire.

   ○ Check whether user forums and communities are available that are associated with free tools and platforms, and see what problems people are talking about. Some free tools have active communities of users and developers who resolve problems quickly; others have less community participation or may be slower to respond to questions about bugs and problems.

- **Examples:** If you have identified that your community mostly wants to have one open-ended conversation online and share a few documents, then free, hosted systems like Yahoo! Groups (http://www.yahoogroups.com), Google Groups (http://www.googlegroups.com), Wikispaces (http://www.wikispaces.com), or Quicktopic (http://www.quicktopic.com) are possible candidates. Note that some people find these tools easy to use, while others experience difficulties and barriers, demonstrating that there are diverse user experiences. Don't assume that your experience will be the same experience that your members have. You may want an online community platform focused on discussions, which allow public or private spaces, and the addition of multiple subspaces. (For an up-to-date listing of forum software, see David Woolley's ThinkofIt site.[1]) Augment a group like this with teleconferencing—available for free through such offerings as FreeConferenceCall (http://www.freeconferencecall.com) or Skype (http://www.skype.com)—and you may have what you need for your community at this point in its development. If your community starts to grow and you need publishing tools, you may consider adopting a hosted blogging tool, such as Blogger (http://www.blogger.com), WordPress (http://www.wordpress.com) or Edublogs (http://www.edublogs.org).

Strategy 3. Build on an enterprise platform

Communities may have access to enterprise-level portals, tools designed for virtual teams, and collaboration platforms. By themselves, enterprise-level systems generally are expensive and complex—overkill when used only for a small community of practice. But they can provide rich resources for communities that exist within organizations deploying these systems. Communities can be satisfied using platforms that were developed for virtual teaming, "learning management systems," content management systems, and large collaboration software packages. These are useful solutions if your community's activities are supported by the tools and features offered in the system, or when you need to support multiple communities. The selection work in this case is to determine whether

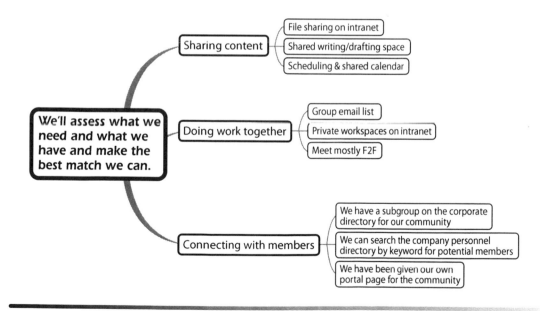

Figure 8.5. Build on an enterprise platform

---

1. David Woolley's Thinkofit, http://thinkofit.com/webconf/forumsoft.html

the existing platform offers the tools needed by a community's orientation. This strategy, illustrated in Figure 8.5, may require a close relationship with the sponsoring IT department.

- **Pros:** Using existing enterprise platforms may be encouraged by a community's IT department to avoid having to provide a separate set of tools. You may not need a special technology budget. Building on an enterprise platform can provide visibility to communities within the sponsoring organization. For content-oriented communities, these systems usually have good document management facilities.

- **Cons:** The collaboration facilities may not be adequate because supporting communities of practice is not the focus of the system. You may need to adjust the facilities for community use or ask the IT department to provide some customizations, especially for conversation-oriented communities with access to a publishing-oriented system.

- **Tips:**

  - Invite the IT department's support early on. Find out what level of support is available for deployment and customization for your community.

  - Use the platform's tools and features that people are most familiar with first, and then add complexity.

  - Understand any back-end compatibility questions. Can new pieces be integrated into existing user databases?

  - Consider mixing and matching enterprise tools with other tools acquired using strategies 1 and 2.

  - If your community spans beyond the organization, make sure the sponsoring organization will allow access by members outside the hosting organization. Beware of firewall issues across organizations.

  - Consider that most enterprise platforms (as well as commercial platforms in strategy 4) are secure from the outside world by design. If your community needs or has a public face, can it exist on the platform, or will you need a separate, public website or other tool?

- **Examples:** Enterprise platforms are increasingly including collaboration tools. LiveLink[2] includes explicit facilities for communities of practice, something that the vendor advertises as a focus. Users of Lotus Notes are creating QuickPlace[3] templates that allow new communities to be set up quickly. SharePoint[4] is available to many Microsoft users. If your organization has a site license for virtual team products

---

2. OpenText, http://www.opentext.com
3. QuickPlace, http://www-142.ibm.com/software/sw-lotus/quickplace
4. SharePoint, http://www.microsoft.com/sharepoint/default.mspx

such as eRoom[5] or Groove[6], you may consider adopting these platforms to help start communities. In higher education, systems like Blackboard[7] and Moodle[8] are increasingly used as enterprise software. Their community features are starting to be used by the institution's communities, even though they were intended to support courses.

Strategy 4. Get a commercial platform

Choosing a commercial community platform can go a long way toward meeting many of a community's needs all at once. This strategy is useful when communities prefer a "one-stop shop" or a community does not have a clear sense of its orientations and wants

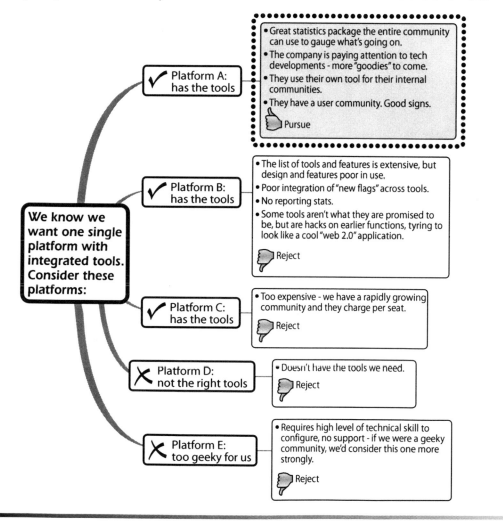

Figure 8.6. Five possible platforms for a community

5. eRoom, http://software.emc.com/microsites/eRoom/indcx.jsp
6. Groove, http://office.microsoft.com/en-us/groove/FX100487641033.aspx
7. Blackboard, http://www.blackboard.com/us/index.Bb
8. Moodle, http://www.moodle.com

a variety of tool options. Some platforms are developed specifically for communities of practice, some do have particular strengths in individual orientations, and others are designed more generally for online communities.

A separate platform can give a community a distinct sense of identity but introduces the issue of how the platform will be present in members' daily lives. It may compete with the other technologies they use. They won't "live" on it as they might on an enterprise platform.

In selecting a community platform, the task is to first identify possible platforms, and then look closer to see if they meet your community's needs as outlined in this chapter. Figure 8.6 gives an example of how a community chose between five possible platforms. First they identified their needs, found possible candidates, and then identified the strengths and weaknesses of each alternative.

Bringing it under one roof

The Learning Systems team at CA, Inc. is a globally distributed team with both internal employees and external partners. Simplification is a key mission for CA, as diverse participants multitask while needing ways to rapidly learn new skills and share information. To focus their time and attention, a single-signon solution with all activity "under one roof" was essential. CA selected eCampus to meet this need for integration and focus. The platform provides specialized training areas for subgroups, discussion threads and RSS feeds on eLearning tools the team is using, thought leadership blogs that highlight what's important with roadmaps for the future, wikis that support best practices in learning and technology, and team-building tools like weekly check-in rooms that cater to the need for social interaction.

- **Pros:** A vendor's focus on the specific needs of communities of practice or online communities in general can bring knowledge and experience with the platform itself. Some vendors support a community of users: tech stewards who share many of your challenges and from whom you can learn about technical details and development strategy. By bundling a set of commonly required tools, a platform can offer a quick start for your community. A lot of the burden of worrying about technology is on someone else's shoulders.

- **Cons:** Commercial systems can be expensive. Several different business models exist, which makes some more expensive in some settings than in others (see Chapter 6). IT departments often resist hosted options as they are outside organizational firewalls and security functions. If used on internal systems, platforms require back-end integration. Some of the vendors are small and do not have a lot of resources to add the features

Having a single platform is not only convenient for group members, but allows for overall technology stewarding through adjustments to the permission-structures. Participants see what is relevant to them, and features such as "what's new" and "what's hot" draw immediate attention to the newest and most active content. The platform allows convenient participation through email, while allowing the team to build a web-based knowledge repository that's complete, searchable, and accessible. The ability to tweak the platform to meet the team's needs, without asking each individual to do any tweaking themselves, helps them focus on their work and learning.

Chapter 8. Technology acquisition strategies

you may find you need. Some of the companies will go out of business. A platform can lock a community into a toolset that may or may not grow with the community.

- **Tips:**
    - ○ Carefully consider the list of tools offered by the vendor to see if your community will really find them usable. Will these tools be adopted by your community?

    - ○ Develop a relationship with your vendor and their user community to see how you can help them with their future development roadmap.

    - ○ Because a platform is usually well integrated, it can be tempting to automatically use the default setup. Consider how you can configure the platform to best meet your community's needs.

    - ○ Because a platform has a distinct identity, consider how it will be present in a member's life. For instance, we have found it important to have flexible email and/or RSS subscription features that help bring people back to the discussions because many people still live in their email inbox.

- **Examples:** Companies that focus on communities of practice include CommunityZero, iCohere, LearningTimes[9], Q2 Learning (eCommunity), and Tomoye (Simplify and Ecco). Some online collaboration companies come close without claiming such focus, for example, ClearSpace[10], Community Server[11], DiscusWare[12], SiteScape[13], WebSphere[14], and Web Crossing. These community systems tend to have a particular emphasis on one or more orientations. For example, some are more content focused while others are more conversation focused.

Strategy 5. Build your own

Some communities or companies have the expertise and willingness to create their own tools and platforms from scratch (see strategy 6 for extending or building on top of open-source software). Building a tool at the same time that you are building a community can be a very creative process but carries some significant risks. This strategy implies a close relationship with a developer and a deliberate investment of significant time and resources.

- **Pros:** This strategy allows you to focus on the specific situation of your community and to develop the system in response to its evolving needs. It may also offer your community something unique and useful that it could not find elsewhere. Where programming costs are lower (using students or outsourcing for programming needs), this strategy can be effective.

9. Learningtimes, http://www.learningtimes.com
10. Clearspace, http://www.jivesoftware.com/products/clearspace
11. Community Server, http://communityserver.org
12. DiscusWare, http://www.discusware.com/index.html
13. SiteScape, http://sitescape.com
14. Websphere, http://www-306.ibm.com/software/websphere

- **Cons:** This path is not simple or easy. Building your own is often more expensive and difficult than you could imagine initially. Be careful not to spend time and resources reinventing the wheel. Even when you work closely with a developer, your community's perspective may not be reflected in their work. Developers often focus on the functions, not on practices of how people use those functions.

- **Tips:**

  o Explore the market and then honestly assess whether your needs are unique enough to warrant building your own platform. Is extending or customizing existing tools or platforms an option?

  o Determine whether building a platform is part of your community's learning and/or community building processes.

  o Have a contingency budget if the project exceeds its initial scope.

  o Consider whether a unique tool will isolate your community from other communities. Do your members belong to many different communities, and will this uniqueness be a barrier to participation? Will your community use this platform exclusively, or does it have to be compatible with other tools and platforms so that building compatibility into the tool will be an additional piece of work?

- **Examples:** Wikis were created to serve a programming community, a great example of the power of "build your own." The Caterpillar Corporation[15] has taken this path with good results, to the point where their platform is now an offering available on the market.

> **Software and community development combined**
>
> In the late 1990s, The Center for Technology in Education at Johns Hopkins University began to obtain grants to develop communities for school teachers which included using online tools. After trying to combine several existing tools, they decided to develop their own platform using Cold Fusion as a development environment.
>
> They had a unique environment and context to serve: supporting school-related communities. They had both a relationship with school districts and a research focus, giving them credibility and trust, which would provide a good test bed for a new platform. A series of federal and other grants to fund development and access to a pool of student programmers provided the resources. So, development made sense.
>
> The Johns Hopkins Electronic Learning Communities included unique ways of reducing teacher isolation in the classroom and supporting their need for sharing resources while meeting the special technical requirements found in schools. They discovered that one of the most heavily used tools was also one of the simplest: instant messaging. But the instant messaging they provided was "school-safe" in that it met teachers' needs to send messages to each other while avoiding the school's prohibition of external instant messaging applications.
>
> http://cte.jhu.edu/index.html

---

15. Sue Todd "Communities of practice at Caterpillar: knowledge is power," Corporate University Journal, Issue No 1, Winter (2006), http://www.corpu.com/newsletter06/cat.asp (accessed February 14, 2007).

Strategy 6. Use open-source software

A possibility that may be present in all the other strategies is that the tools selected are open-source[16] (OS) software. Open-source software is free but often requires customization. Some communities want to support the open-source movement to make software accessible to more people. They may embrace the ability to take a base set of code, customize it for themselves, and contribute this work back to the larger open-source community. Figure 8.7 gives an example of a community that selected open-source software.

Many open-source projects are communities of practice themselves, making them allies in some way with other communities. If open source is critical to your community, it might be a driver in your technology selection. Open source platforms share many of the characteristics of commercial platforms, so refer back to strategy 4 as well.

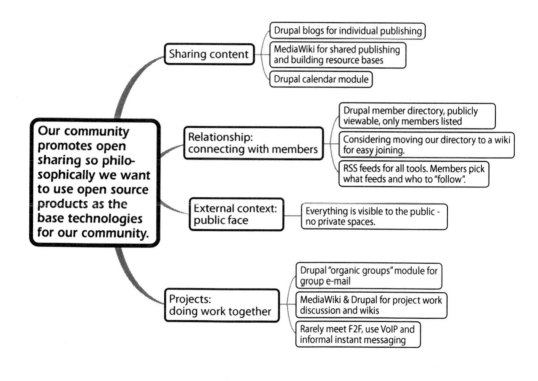

Figure 8.7. An open-source selection example

---

16. From Wikipedia: "Open-source software is computer software whose source code is available under a copyright license that permits users to study, change, and improve the software, and to redistribute it in modified or unmodified form. It is the most prominent example of open source development." Wikipedia contributors, "Open source software," Wikipedia, The Free Encyclopedia, http://en.wikipedia.org/w/index.php?title=Open_source_software&oldid=225588876 (accessed July 15, 2008).

This strategy can imply a close relationship with the open-source community surrounding your chosen product. Open-source systems range from very stable, ready-to-use systems that are equivalent to commercial platforms, to systems that are in the process of being proven, where you need to be a community member to use them effectively. Open source can also require a close relationship with a developer to customize the software for your community.

- **Pros:** The variety and quality of open-source collaboration tools are increasing. It is a vibrant market sector worthy of attention. Open source usually brings along the great resource of an active user community that is the heart of open-source development. As more communities adopt OS tools, the tools grow and mature.

- **Cons:** At the writing of this book, much open-source software still requires substantial programming to install and configure. So while the code is free, configuration and hosting can carry significant costs. Not all open-source products are mature and stable.

- **Tips:**

  - Evaluate the product's maturity. Can it be used without any modification at all? Has it gone through more than one version? Determine which version is most stable and appropriate for your community.

  - Consider configuration costs. You may have to spend significant funds to configure and support the software. It is important to make these choices with enough information about both long- and short-term implications.

  - Observe the community around the product. Is it vibrant and productive? Do clear development maps and decision making processes exist? Are security holes identified and plugged quickly?

  - Are commercial versions with additional features available? Could those versions be of use to you? Also, some open-source tools are available as hosted offerings for a fee so that you don't have to worry about maintenance or upgrades. A blended approach would be to choose a free open-source tool but pay for additional features or hosting.

- **Examples:** Drupal[17], Joomla[18], and Plone[19] are open-source platforms that provide interesting community-oriented characteristics but require some programming expertise to configure and deploy them. Moodle[20] is an open-source online course management system that has conversation-oriented facilities and has been used by commu-

---

17. Drupal, http://www.drupal.org
18. Joomla, http://www.joomla.org
19. Plone, http://plone.org
20. Moodle, http://www.moodle.org

Chapter 8. Technology acquisition strategies

nities as a home platform. WordPress is an open-source tool for blogging. Take a look at SourceForge[21], an online repository for open-source tools available and under development.

Strategy 7. Patch elements together

Advances in technology, often identified under the Web 2.0 banner, are creating new opportunities for communities to construct their own platforms by patching together easily available pieces, sometimes even with no programming skills. This idea of "small pieces, loosely joined"[22] helps solve one of the growing problems associated with the "use free things" strategy: not everyone is adept at jumping from tool to tool. Some people seek more integration. The patch-work strategy provides some level of integration among tools by "stitching" together elements from different sources to suit a community and its members. Some key technologies that make this possible include:

Mashup for a community map

CommunitiesConnect is a community of non-profits in the social services sector who are interested in the use of technology to achieve their goals. They use an API from Google Maps to create a custom map of their members that is part of their community directory, automatically locating each of the member organizations on a map of the state of Washington. The map is integrated onto an open source Plone platform that is used for basic information sharing.

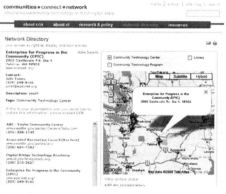

http://www.communitiesconnect.org/network-directory

- **RSS feeds** deliver the ability to easily pull in content from other sources into a community's online platform. For example, a community can bring in all the blog posts from members' individual blogs, regardless of the platform where these blogs are hosted.

- **Open application programming interfaces, or APIs** offer ways to combine tools and features by making the functionality accessible beyond the original designers. For example, Facebook has provided an open API to its social networking platform, enabling other sites to create tools that work on Facebook. (A wiki provider could create a way to have its wiki appear on Facebook, in this case). This then offers communities a chance to use not only Facebook but other third-party tools, which can be integrated easily on a Facebook page with no programming skill needed.

---

21. SourceForge, http://sourceforge.net
22. Tim O'Reilly, "Open Data: small pieces loosely joined", O'Reilly Radar, September 4, 2006, http://radar.oreilly.com/archives/2006/09/open-data-small-pieces-loosely.html (accessed February 14, 2007).

- **Widgets**[23] are tools made by third parties that are offered for integration into a site, usually simply by copying and pasting code from the provider to the community platform. For example, a community can integrate an external calendar into its platform to extend its functionality.

- **Mashups** take the idea of patching data sources or functionality from more than one location to create something entirely new. The most common example is connecting data to maps, a typical mashup called geomapping. From a community perspective, this is usually about combining community data with some external data in a way that has value to the community.

Currently this strategy still requires some technical knowledge, but it's becoming much easier. This makes the patching strategy one to watch as both the technology and creativity of applications advance.

- **Pros:** This strategy allows a community to add and subtract tools and content, and readjust features incrementally over time. Even when one platform serves as a base for the community, new functionality can be added or removed with minimum effort. Such an approach potentially increases the opportunities for members to creatively steward technology as the evolution of tools and standards makes it possible for people with less technical backgrounds to do this work. It can allow gradual evolution without the disruption of a migration from one platform to a completely different one.

- **Cons:** Easy-to-implement enhancements may seem "cool" to the technology savvy but won't always make sense to the full community or work on a larger scale. The fact that change is easy to introduce can tempt anyone to introduce new widgets casually, ignoring members who find change irritating. Integrating a third-party hosted widget may not work for a large community. For example, the widget maker may withdraw hosting or support for the widget with no notice. Third-party tools may carry viruses, Trojan horses, or other malware that could cause significant damage.

- **Tips:**
  - Test new feeds, mashups and widgets off the community's main page or site to avoid distracting those with little time or attention, while still giving others the chance to experiment. Once a new innovation is stable and useful, move it into the main community view.

  - Remove feeds and widgets that fall into disuse.

- **Examples:** The effort and technical expertise needed to patch things together vary quite a bit. At the easy end of the spectrum would be adding a Flickr photo widget

---

23. For examples, see Quick Online Tips, "The Great Flickr Tools Collection, March 29, 2005, http://www.quickonlinetips.com/archives/2005/03/great-flickr-tools-collection (accessed February 14, 2007).

to a community page, so that recent community pictures are rotated; this can be done with very little technical expertise. Patching can require more sophistication—for instance, integrating a traditional discussion forum such as phpbb[24] and Google maps to tag member locations[25], or linking event calendars and maps to facilitate face-to-face meetings in geographic subgroups.

Each strategy discussed in this chapter takes a slightly different approach to acquiring software for your community. Each one can build upon community circumstances and needs and has implications for your budget, resource availability, and your comfort with technology. As you make incremental changes to your community's digital habitat, here are a few tips.

- **Start with the simplest, least expensive solution that you think will work.** Identify tools already in use, or available. Determine whether you need to buy anything all! Justify the decision to buy technology with real community experience.

- **Learn from other communities and their tech stewards.** Once you've picked a strategy, find others who have used that strategy and talk with them.

- **Develop a testing plan.** Depending on your strategy, find a way to include your community in testing the technology. Ask a vendor for a test space. If they decline your request, beware. Use free tools for small, experimental activities with the community.

- **Once you recover from the first round, keep an eye out for what is next.** This is a cyclical process and your strategy may change in the next cycle. Evaluate what you see based on what you learn about your community's needs and the experience of other communities. Prepare for the next cycle of decision-making.

---

24. PHP Bulletin Board, http://www.phpbb.com
25. All Things Beer Map, http://www.nhbrewers.com/mapbeer.html?lat=40.71463&lng=-74.005806&zoom=11

Stewarding technology in use 9

The focus of this chapter is the ongoing role of tech stewards in the daily life of a community, including the process of implementing new technologies. Offering a mix of guidelines and specific suggestions, we focus on your practical activities and responsibilities, highlighting the creative, inventive, and down-in-the-trenches work of technology stewarding. The role of the tech steward is not just to manage the configuration, but to make it a productive habitat.

Stewarding is a diverse practice, so we start with a few principles that we hold when we are playing the role:

1. **Keep the vision of your community's success above the technical details of technology implementation.** Participate in the community's life as much as you can. Maintain the vision and use your experience to guide your goals or performance criteria, while maintaining some flexibility with uncertainties and changes. This sets tech stewards apart from technologists who focus primarily on technology support.

2. **Keep the technology as simple as possible for the community while meeting its needs.** That way, the technology is shaped by the community at the same time it is helping to shape the community. No matter what goodies you find in your scanning of the technology landscape, focus on the simplest structure you can get away with at the beginning. Use community successes and failures as keys to subsequent steps. From there you can incrementally add, subtract, and adjust your tools based on their use by the community.

3. **Let the configuration of technologies evolve as the community evolves.** Members may stop using some tools and slip in others they prefer. Tech stewards may be unaware of tools that members are using that are not part of the "official" configuration, but through use, have become part of it. A community's use of technology becomes more intricate over time. This attention to the overall configuration is about technology in use—how the tools and the community's practice fit together eventually.

4. **Use all the knowledge around you.** People in your network may be your richest resource, with perspectives that can inform your work of scanning, choosing, imple-

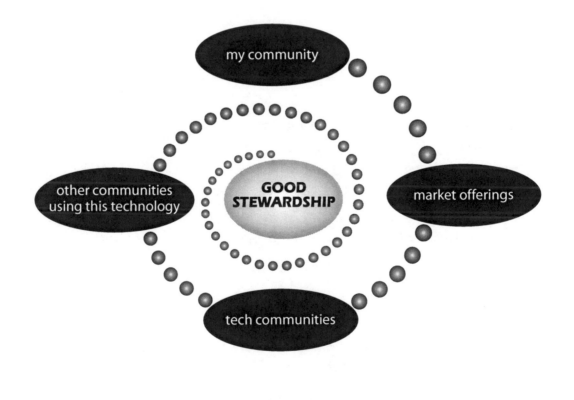

Figure 9.1. Sources of knowledge for good stewardship

Chapter 9. Stewarding technology in use

menting, and supporting technology in use. Put your community to work to reduce the uncertainty about what you don't know. Consider these factors:

- Your own community has knowledge about local technology conditions, for example, and it has skills that can be used in your stewarding effort if you communicate what's needed.

- Other tech stewards are key sources of knowledge, especially those who support similar technologies or orientations. They may suggest technology they like or avoid, and share their practical experience.

- Product-focused user communities or technology-oriented communities of practice provide valuable context and advice, often giving better technical support, with more context than vendors can provide.

- Working with vendors and external technologists, you can bridge their technical expertise and your knowledge of the community.

To tap this network of resources, you have to connect and participate. Share what you learn as a tech steward with others.

5. **Finally, "back it up."** This may sound obvious, but the data brought in and accumulated by a community is important, including the membership list, shared resources, and artifacts of your interactions. In many organizations you can take for granted that the IT department keeps reliable backups. However, if the community's key data is all on one member's hard drive and that drive crashes, or if an application service provider has a catastrophic failure and its backup system doesn't have enough redundancy, your community will sustain a serious loss. Make and store copies of key files appropriately.

Stewarding in the foreground is different than in the background

We can create an idealized version of "what a tech steward should do." But in reality there are always time and resource limits. This makes it important to focus on what matters most to your community—where it is right now and where it might go in the near future. To help you think about the specifics of where you are right now, we have organized information based on the amount of focus on a particular activity. As you scan these, you can consider where you might invest your current stewarding energies.

In some cases, when a community faces significant changes in technology or technology-related practices, technology is in the foreground—and tech stewarding has a central role with high visibility. Two important examples are:

1. Implementing and deploying a community platform or set of technologies (new or migration)

2. Community closure and end-of-life issues for distributed communities

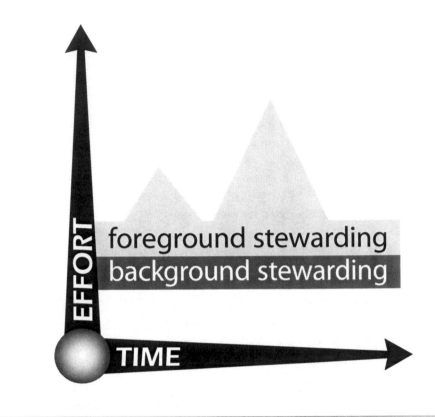

Figure 9.2. Foreground and background stewarding effort over time

In most cases, technology is more or less "in the background." This occurs in day-to-day situations, where technology tasks are not the central focus of the community, and stewarding is less noticeable: a set of tasks that one or two people attend to with little visibility. In these situations, the goal is for the community's attention to be on its business, not on its technology. These stewarding activities include:

1. Supporting new members in their use of the community's technology

2. Identifying and spreading good technology practices

3. Supporting community experimentation

4. Attending to community boundaries created by technology

5. Assuring continuity across technology disruptions

Chapter 9. Stewarding technology in use

These are some of the many issues that tech stewards should pay attention to, even when the rest of the community is not interested. Obviously, a distributed community that depends on technologies to learn together must always pay some degree of attention to its infrastructure. In practice, the amount of conscious attention given to a community's technology varies naturally, over time.

Major transitions: technology stewarding in the foreground

As stewarding moves to the foreground during major transitions, the community pays attention to its technologies—and consequently to its tech stewards. The tech steward's relationships become more intense. Tech stewards will encounter anything from rapid, positive adoption, to grudging acceptance, to proposals for completely different directions, to outright rejection. Tech stewards may find themselves enjoying or disliking the attention.

1. Implementation and deployment of a community platform (new or migration)

A major event in the life of a community is the deployment of new technologies or platforms. This could involve developing a community's first formal infrastructure or the migration of an existing community from one platform to another. If the community is just coming into existence on a new platform, there are more unknowns and it can take a period of

Stewarding prompted by community needs

A global community had incrementally added technology to their configuration over the years, mainly through the generosity of its sponsor and with little attention from the community. The members started with an email list and added a content management system for their website.

Growth of the community and a change in sponsorship led the members to focus attention on a potential change in platform. Their initial web-based content management system did not have a robust user community and was not easy to configure. The loss of programming expertise to support the platform was a worry.

A subgroup was formed to review community needs, make a recommendation for a new platform, and seek new organizational sponsorship to support the move. This group continued to take an active and visible role in stewarding the community's technology, even after their original task was complete.

sustained use before a configuration feels like "home." If the community is migrating, then the new platform may enlarge its membership, enable new activities, and even change its orientations. It may also rub up against existing community habits and vested interests. We talk about these various situations together because of their similarities with regard to technology stewardship: both make it visible and make the work more challenging.

For this section, we focus on the platform level, but the ideas can apply to any major change in a community's technology configuration. Implementation of a new platform highlights the interactions between communities and technology. The change affects how the community works and experiences itself.

Whether a community is being launched on a new platform or is migrating from one platform to another, consider the "before" and "after" in community terms. The change your community will experience is both technical and social. It requires both a project management component to organize the technology implementation and the social strategy to address the members' experience. Social change usually takes much longer than we think. Managing the change could involve looking into member skills and readiness, as well as considering how relationships in the community could be affected. How much time will someone spend learning to use new technology in order to participate in a community that is new and still tentative? For migration, a new tool or platform may not replace everything that was there before, or be "everything to all members." For instance, if a community moves from email to a website, members will most likely still use email, but will have to learn to use it for different purposes. The process of migration can easily leave some members behind, either as resisters or as people who have difficulty learning new tools and practices. These challenges require you to leverage all the technical and community resources you can muster.

Planning pointers:

- Pick the right scale of planning and change management. Ensure that your plans for technical implementation serve the community's need for a home and don't overshadow the community itself.

- Working with the community, lay out clear goals, expectations, and commitments. "By the community, for the community" is a good motto at this point. Create a roadmap with your specifications, timelines, and other pointers that make sense both to the community and to any IT partners who might be involved.

- Identify others who have used the technology you are implementing and find out what you can learn from their experiences. If you can join a user community related to your new platform, do so.

- Identify a network of early adopters within your community to help think through and experiment with the possible implications of new choices and designs. Early adopters can be invaluable in helping integrate new tools into the configuration.

Technology pointers:

- Identify the technology partners with whom you need to coordinate, and determine who has a decision-making role.

- Coordinate all the partners involved in the technology implementation. This could involve signing up for a hosted service (an ASP), collaborating with a developer who will configure or install software, or coordinating with people in your IT department. Figure out who you need to talk to and talk to them early on.

- If you are migrating from a legacy system, look carefully at issues that can range from data transfer to system integrity. If relevant, talk with the IT department and understand what they do and do not support (and what they might contribute).

- Consider the scope of decisions. Much of the struggle in a big implementation project involves tradeoffs between short-term and longer-term goals. For example, if your community platform is on its own server and therefore has its own URL—especially if your members will send and receive email—allow enough time to name it and get its name registered. Think carefully about the URL and what it will mean (to community members and non-members as well), as that's hard to change later.

Practice pointers:

- Use small experiments that can be abandoned at low cost. Especially when you are adding a new technology that is not a sure thing, try to understand its implications in practice for your community. Consider how it works with existing tools people already use. Invest enough effort to really find out how a platform works and how it is likely to affect your community, while allowing the community to abandon it if it turns out to be a failure.

- When a community uses a tool, the shared assumptions about how to use it constitute a practice. During implementation of a new system, provide some process for the development of new practices, such as using part of a community meeting to look at the tool and compare how people are using it. A little sharing can go a long way. Too much can swamp the conversation.

- It is often a good idea to make a new platform feel "lived in" at launch time to demonstrate new functionality and benefits and to give it a community feel. Initiate interactions in discussion spaces, pre-populate a wiki, or write some initial blog posts.

- When making the transition to a new platform, determine whether you will need to move files, member databases, discussions, or other historical artifacts from an earlier platform or from several dispersed tools.

- In a new community, members may have relevant materials to contribute right away. You will need a structure to organize rapid growth. For communities that focus on creating and sharing materials, consider taxonomies for complex sets of materials.

- During the transition, develop a "crosswalk" to help people find the new location of community content and tools they were using on the old platform. For example, create a table with links for community materials on both platforms, particularly if the two platforms co-exist for a while. Note what gets left behind in the move.

- If convenient, launch new tools and platforms at regular community events. For example, a face-to-face meeting is a good opportunity to demonstrate a platform.

- Leverage existing channels of communication. Communities of practice typically use multiple channels of communication. Use any existing resources such as email lists, regular meetings, or whatever other channels of communication exist for your community to help make platform configuration and adoption reflect the community's wishes and needs.

2. Community closure and end-of-life issues

As a community comes to the end of its lifecycle, tech stewards may need to archive and preserve a community's history before the "lights are turned off." Member profiles and account information, the data they've left and the artifacts they've created, can persist (and have real value) after the community ceases to exist or is in a dormant phase.

- Work with the community to identify when technology services should cease, what artifacts should be preserved, and what should be deleted.

- Preserve identified community artifacts. If necessary, find a safe haven or location for a community time capsule that contains key community artifacts. Do other communities exist that may have an interest in being the caretakers of a legacy left by a defunct community?

- Shut down online spaces or cancel contracted services with ASPs.

- If privacy agreements dictate, erase sensitive materials.

Carrying key artifacts forward

When the original Electric Minds community was dismantled due to the end of its original funding, some members wanted to keep archives of important community conversations. Others wanted to retain the powerful images associated with the community. Individual members downloaded discussions and images, and then posted them on their personal websites. These were later linked to and referenced by the next iteration of the community after it moved to a new platform and URL.

http://www.electricminds.org

http://www.abbedon.com/electricminds/html/home.html

Ongoing concerns: technology stewarding moving into the background

Technology stewarding often moves to the background after a major transition. From a community's perspective, it is appropriate to keep technology somewhat in the background. Members have a limited amount of time, and, even during periods of change, usually have limited interest in technology for its own sake. Do what you can to avoid "change fatigue." Consider how your community's technology configuration is evolving incrementally, ensuring it is moving as fast as the community needs it to evolve and no faster. Tech stewards have a limited amount of community attention they can call upon, so use broadcast communications judiciously. If you do too much broadcasting, people start ignoring your messages.

Moving to the background does not imply that the job is over, or that no effort is involved. Tech stewards are attending to the daily technology heartbeat of the community. This may take very little attention or a lot, depending on the community's vitality. Good practice involves attending to the areas described below.

1. Supporting new members in their use of the community's technology

New members of a community face both social and technical hurdles. In many communities, registration and orientation are two important and ongoing technology stewarding activities.

Some tools require members to register, while others (for example, telephone bridges) typically do not. While self-registration is possible and convenient, many communities choose not to fully automate a part of the registration process so they can approve membership requests. They often blend registration with a welcoming process, and include an orientation to the basics of a community's tools and other social agreements.

Learning to use a new tool

A network of community service providers had banded together to learn more about how to fundraise effectively. They needed a simple platform to share materials and chose a wiki. However, most of the members had no experience with wikis. So they set up a monthly phone call to walk members through the basic features and practices for using the wiki. Everyone on the call watched the call leader use the features, and then they tried it themselves.

At the end of each call, they did a quick review to find out what value people got from it. The first thing they said was, "It was great to meet new members and hear what they were doing." Learning about the wiki was secondary. From this feedback they concluded that learning about new tools needed to be combined with learning that really motivated community members. Few were interested in the technology itself, even though it was a prerequisite for the community's life together.

Practice pointers:

- How much do you want to integrate orientation into a community's registration process? Small communities may want orientation to be a very personal, intimate experience. Large communities with very open boundaries may wish to make it fully automated.

- Consider a new member's need to learn to use the community's configuration. What is the minimum they need to know to begin participating? How can they learn the rest? Don't try to teach them everything at once or expect instant proficiency across the board, particularly in complex configurations. Enabling initial participation quickly is the key principle.

- The process of onboarding new members is usually a mix of a formal design with an informal execution. If you have a large community, engage other members to help support both the formal and the informal. One-on-one support for new members is particularly useful.

Technology pointers:

- Consider how the design of your member registration process supports the social aspect of joining the community. Does the current software configuration for handling membership requests meet the community's needs?

- Develop prompts to help new and existing members add their personal information to the member directory, if you have one.

2. Identifying and spreading good technology practices

Communities learn how to use technologies in formal and informal ways. While the adoption of new tools sometimes requires formal training, informal sharing of practices often has an even bigger impact because of its ongoing nature. Word of mouth, one-on-one assistance, and experience help members use shared tools more effectively. Technology can also be used, for example through frequently asked questions (FAQs) and contextual help features found in some platforms.

In supporting the formal and the informal use of technology, tech stewards offer a special set of eyes and ears, which allows them to spread the word about what is learned. We have noted repeatedly that the way communities use a tool can be different from what a designer intended, whether the community is aware of it or not. It is useful to find out how tools are really used. Be curious about how and how much a community is using a specific tool. Because for the most part we experience technologies individually, even tech stewards sometimes have no idea what other members are experiencing. Everyday tools, like an instant messenger, can be used in many ways. For example, some people leave instant messages for recipients who are not online, knowing that the message will catch their attention when they return. Community provides a context of experience where people can notice each other's uses and talk about it. Seeing how members use everyday tools can stimulate important lessons about technology that will benefit the whole community.

> **One question uncovers many practices**
>
> When they were first launching, a small, international community of practice decided that welcoming new members could be a real community-building activity. To keep new members engaged, everyone was sent a welcoming email telling them they had been automatically subscribed to the two main community discussions. This worked well for many members, but after the message volume increased, some people felt overwhelmed and wanted to unsubscribe. Although there was an "unsubscribe" link in every message, not everyone saw it or understood how to do it. When a member posted their frustration, others responded with their strategies for managing messages and unsubscribing to discussions. These strategies were far more diverse than the basic approach offered in the welcome email. As a result, a new FAQ was developed to help future members have a better sense of how to manage their participation in the email discussions.

Chapter 9. Stewarding technology in use

Practice pointers:

- Make the community's "tool use" visible. Encourage situations where community members can observe each other using common tools. Make the use of tools a legitimate topic of conversation, but keep it in the context of the work of the community. As members of a community, tech stewards can occasionally change the subject of community conversations, bringing technology to the foreground.

- Leverage community activities that spread technology skills. Consider how certain events or transactions could highlight the utility of the technology being introduced or make its adoption easier. Face-to-face events are good opportunities for spreading technology skills, because people can observe each other, but they also have their limitations (for example, time pressure, or being away from the normal workplace).

> **Tech stewards look over each others' shoulders**
>
> At an international conference on e-Learning, one of the presenters had demonstrated the use of Wikispaces, del.ico.us, and Twitter in her communities during her presentation. Later, she happened to be sitting next to another community steward and noticed that he was creating a Wikispaces page for his community. He had been inspired in the moment and had jumped in to experiment.
>
> He leaned over and said, "Tell me more about this del.ico.us thing! I liked Wikispaces, so I figure I should pay attention to your other recommendations!" Together they swapped ideas, showed examples from each other's communities and ended up smarter about the tools, just by looking over each others' shoulders.

- Pay attention to the way different tools give access to people or exclude people in the community. Also, people's beliefs about how specific tools are "good for" specific purposes and used individually or with each other make a big difference. These factors influence whether and how people adopt new tools. Capture and share anecdotes and examples of tool usage and adoption.

- As new techniques to make tools work together are invented, surface necessary agreements and practices, and consider how that new knowledge can be held and spread.

- Try to develop some competence using all the tools that are included in the community's tool configuration, regardless of your personal preferences. Awareness of all the tools in use helps a tech steward understand the way tools work together.

Technology pointers:

- Use monitoring tools and features. Most tools produce data records of some sort as a by-product of their use. Now you can even integrate external analytical tools such as Google Analytics[1] into platforms, extending the amount and quality of data available.

---

1. Google Analytics, http://www.google.com/analytics

That data can help tech stewards assess tool usage or community activity patterns. Data that tracks tool usage can help you ground your observations with objective measures. For example, some web-based discussion tools provide data on page views in different topic areas or indicate who has read what. Some telephone bridge services report the connect time for each participant by phone number. These tools will always constrain what you can "see"—both technically and socially—but they can raise new questions or spark new insights about a community's activities and patterns.

- Designate a primary location to store local tips and instructions about all the tools. Work to make it universally accessible and known to everyone in the community.

3. Supporting community experimentation

A community's configuration will evolve, often through experimentation. In collaboration with the community, tech stewards can help transform experiments, accidents, or local discoveries into community-wide practices and agreements that advance the community's capacity. This experimentation must be balanced with attention to reasonable preservation of the infrastructure.

Experimenting as the heart of a community

Some communities experiment with new technologies. The EdTech Talk community does it all the time as a core activity. This loose-knit community, consisting mostly of teachers and school IT directors, is interested in the application of technology in education. One of the twelve shows on the network, Edtechweekly, meets once a week for a live, web-based event to explore new technologies and how they can be used. Between 15 and 30 people usually participate. People send in suggestions during the week through delicious.com or just show up with new tools and the Edtechweekly team tags interesting things for discussion that week. The group thinks together, evaluates, and heckles what's been brought forward. The whole process is extremely useful for participants.

But there are issues of scalability in the community's experimentation. Dave Cormier, the community leader, shares a few of their findings as more people come into the community to experiment together. "Success is the worse thing that can happen to you on the net. Failure is fine. If a tool works fine for five people, fine. It gets complicated when other people come to play. Text chat may be great with six people and useless at 150."

By experimenting, the EdTech Talk community continues to figure out how to make its core activity of experimentation scale up.

http://www.edtechtalk.com/About_EdTechTalk

Practice pointers:

- Encourage experimentation with new tools and practices. Support informal and large-scale experiments.

- Remind experimenters about the fundamental purposes and needs of the community.

- Report on the results of all experiments, whether "successful" or not, in an even-handed and inquisitive tone.

Chapter 9. Stewarding technology in use

- Act on the results. If a new technology or practice is promising, support its wider implementation.

- When a tool moves from an experiment to a core technology that's used on a regular basis, treat the transition as an implementation, as discussed above.

4. Attending to community boundaries created by technology

Technology creates a variety of boundaries. Some communities have large peripheries—open to search engines and anybody who drops by. But for communities with member databases, being registered indicates that you are "in the community." Within a community, specific roles (including that of tech steward) may require different levels of system authorization. New subgroups may need specific setups in the software. What is visible to people outside the community, as defined by technology access rights, describes another set of boundaries that are created and implemented with technology.

Practice pointers:

- Some members of a community will have a strong preference for one tool or mode of working (for example, they want to avoid synchronous or asynchronous tools as much as possible).

Darned if you do, darned if you don't

One community set up a wiki to make it easy for its widely distributed members to contribute to a shared knowledge base. To contribute, all anyone had to do was hit the "edit" button. No registration was required. This was a great resource because the content could be shared with everyone's home organization.

After a few months, the wiki started getting spammed with advertising. Page after page had to be "reverted" to the actual content that had been authored by legitimate community members. Reluctantly, the tech stewards closed down the open access, requiring anyone who wanted to edit a page to register. This reduced the number of contributions but eliminated the spam problem.

http://www.km4dev.org/wiki

- Some obscure tools or features may be accessible or even make sense only to members who've participated for a long time.

- Build bridges across these boundaries wherever possible, such as having a page that lists all the community's tools, or cross reference activities that happen in different places.

Technology pointers:

- Have a clear understanding of community roles and how they are reflected in system permissions. Keep track of who should have the ability to do what on a system. Set up a special account or find a method for the express purpose of seeing a platform "as a regular user would see it" since your special privileges may not reveal some of the user problems.

- Identify which, if any, of the community's content and activities should be made publicly viewable and set the appropriate access permissions.

- In cases where specific security practices are required, ensure the software is configured to meet those requirements. If an organization's security practices govern community activities, you may need to learn a lot about the larger security environment.

5. Ensuring continuity across technology disruptions

A community that depends on technology needs to be assured that its infrastructure won't be disrupted or completely disappear. Resources such as email lists, community websites, or libraries can look like stable resources, but they are susceptible to disruptions. People forget that pressing that "one button" can delete entire discussion forums. What looks like a small tweak creates a cascade of problems. Toolmakers introduce changes through bug fixes and new releases. Change, in big and small increments, is part of the community technology experience, but attention must be paid so that the change does not destroy basic community technology functions. Stewarding is both proactive and reactive to these sorts of changes in the community's configuration.

Technology pointers:

- Ensure that the community's core technologies are not disrupted while changes are being made to the configuration. For example, test new tools on a test platform rather than jeopardizing the main platform.

> **A cautionary tale**
>
> One community, the ACT-KM group on Yahoo! Groups, suddenly disappeared one day. The list supported a large and active community, in existence for several years, which started as a face-to-face community with regular conferences in its Australian home base. No one ever expected something like that to happen. There was no current backup of members or email addresses. They could not find out what had happened from Yahoo! Groups. Community leadership sprang into action, alerted members they knew about the group's online disappearance, and began creating an alternate, independent site.
>
> At the same time, they used their network to find someone with contacts within Yahoo, as they were getting no response from Yahoo support. They wanted to find a way to recover their messages and membership data. Those connections yielded results. The community was able to learn why their site was closed down and regain their archives. They chose to stay with their new site, and developed some practices to back up their community data. The transition allowed them to look at their needs, and they expanded to a site with a blog and enhanced their email list.
>
> http://www.actkm.org

- Periodically create or update an inventory of all the tools and platforms in use. Include technologies put in place for the community and those brought in, ad hoc, by members. Some of these may not be obvious; ask members about the tools they're using.

- Clean up old files and remove clutter. Some community sites become so cluttered that they become unusable and unwelcoming.

- It is worth repeating that it is important to ensure regular backups are made. Don't assume anybody else is doing it!

- If the tools have master passwords or system accounts, make sure more than one member of the community can access them as a backup. Keep passwords in a secure place.

- For commercial and open-source products, monitor user communities associated with your tools and platforms to get advanced notification about changes or new features. Pay attention to feedback on early upgrade releases to help you decide when it is time to upgrade.

Practice pointers:

- Periodically review who is responsible for doing what in terms of technology stewarding.

- If a community uses many different tools on different platforms, it is useful to make a regular tour of them to see what is being used and what is falling into disuse. Some tools are out of the tech steward's main view, used only by subgroups. These tools may have undergone some changes or need security updates.

- Clean up member databases. As appropriate, remove people from the membership list or database after they leave the community. In rare cases, members may ask that all the materials they have posted be removed. If this is consistent with a community's agreements, this may be a job for the tech steward.

Beyond cycling between stewarding in the foreground and background, there is the challenge of stewarding in a rapidly changing technology environment. No digital habitat is completely static. Today we may be dealing with Web 2.0 issues and 3.0 tomorrow, but dealing with change is a constant. New tools create new stewarding challenges.

The new technologies point to the users' ownership of their software-mediated experience. They offer new ways to bridge (or separate!) the individual and the group. The tools focus on easy publication, easy formation of groups, and the ability for individuals to drive their use of the tools. In many ways, individuals become central players as tech stewards. So, while a community's tech steward may be looking for ways to create a shared set of tools and experiences, the individual members may be choosing alternative paths that do or do not complement the goals of community stewardship. Your community may share a digital habitat with many other communities.

Stewarding technology involves knowing a lot but it also involves a lot of intuition, guesswork, and the patience to tolerate uncertainty and not knowing. Tech stewards face fundamental questions that can't be answered in advance or from a distance. This uncertainty requires insight and inventiveness on the part of tech stewards and the community, whether through making do with what's available, inventing technical workarounds, or forging ahead with new design efforts. There is always the question of whether a certain tool or configuration is "good enough." Determining what communities will tolerate or demand—including their needs, interests, and motivations—makes stewarding interesting work. This kind of work cannot be reduced to one formula.

Stewarding technology is an emerging and dynamic practice. A good practitioner balances technical and community knowledge, draws on the community's resources, and attends to many different levels of technological change with their potential to shape community evolution.

Action notebook

This chapter is a practitioner-oriented summary of the book so far. It is couched as an "action notebook." With checklists, tables, and questions, it takes you through the steps of stewarding technology and outlines what to keep in mind at each step.

While a summary, this chapter does not follow the order of the book. In the body of the book, chapters are introduced in an order that reflects conceptual prerequisites: which concepts need to be introduced first so that the rest of the text makes sense. Here we revisit what we have said, but it an order that approximates better the way in which these ideas can be put into practice:

- **Preamble:** reflection on the role of tech steward

- **Step 1: understanding your community**, its characteristics, orientation, and current configuration

- **Step 2: providing technology**, choosing a strategy, selecting a solution, and planning the change

- **Step 3: stewarding technology in use**, in the life of the community and at its closing

Of course, the actual work of stewarding technology for community is not quite that linear, but it is still useful to think about it in terms of these steps. For each heading, we provide the reference to the chapter(s) where the material is discussed. We have provided a downloadable document on the book's website. We suggest you use that so you can work with the tables and look at pages together with other community members (http://technologyforcommunities.com/actionnotebook).

Preamble ➡ Being a tech steward

Start with some reflections on your role as a tech steward.

➡ About you as a steward	(Chapter 3)
Personally:	
• Why are you doing this? What do you expect?	
• What is your background (e.g., technology or community) and how does this affect your biases?	
• How much energy and time do you have for stewarding?	
• How will you learn what you need to know?	
• Who can provide support to you personally?	
In your community (For purpose of simplification, we'll use the singular in this chapter, but all that we say also applies to tech stewards who serve multiple communities):	
• What is your relationship to the community?	
• What is your relationship with community leaders?	
• What ways do you have to understand the community activities/practice?	
• What or who gives you the legitimacy to play this role?	
• Who else is interested and could help you by offering resources?	

Tip: Look for labor, financial, and in-kind resources. Don't forget the power of voluntary contributions. They can build the spirit of community.

➡ About the work of stewarding	(Chapter 9)
Check the principles that are particularly relevant to you:	**Why are they relevant? What do these mean to you?**
☐ **Vision before technology:** What is the vision of your community's success? Place this above a list of technical specifications.	
☐ **Keep it simple:** What is the simplest solution for your community at this point in time? Is that "good enough"?	
☐ **Let it evolve:** Are you helping the community have a sense of its own evolution? Think of technology, not as a system, but as integral part of the evolution of your community.	
☐ **Use the knowledge around you:** Who can you tap to learn with and from?	
☐ **Always back it up:** What is your data back-up plan?	

With all this in mind, you are ready to proceed.

Step 1. ➡ Understand your community

No matter what your relationship is to your community—a core member, a leader, a peripheral member, not a member at all—the first and foremost step is to understand your community and its circumstances.

Tip: Consult with your community as you build your picture. It could even be useful to set aside a small amount of time to talk about all the following issues as a group—without distracting the community from its main interests. You may gain greater insights about what members experience, what they aspire to, and what technology they do and do not use or want.

1.1 ➡ Community characteristics　　　　　(Chapter 7)

Lifecycle

Where is your community in its lifecycle?	What you need to focus on:	Special needs
☐ **Just forming:** need basic tools to connect, but not sure from there	Discuss the potential of some basic tools with members, explore what ideas it might give them, and see what they might bring in with them.	
☐ **Self-designing:** in formation stage, but with a strong sense of what it wants to accomplish	Contribute ideas to the design. Analyze systematically the implications of their community design for technology, infrastructure, and technology skills.	
☐ **Growing and restless:** ready to add new functionality to its tool configuration	Try to make this a community reflection and self-design event. Does their restlessness suggest a major change, such as a transition to a new platform?	
☐ **Stable and adapting:** just needing some new tools	How much disruption will the community tolerate? How will the new tools be integrated into or affect existing practices?	

Constitution

☐ **Diversity:** How diverse is your community?	
• What are the different types of members and what are their levels of participation?	
• How spread apart is it in terms of location and time zones?	
• What language(s) do members speak?	
• What other cultural or other diversity aspects may affect your technology choices?	

☐ **Openness:** How connected to the outside world is your community?		
• How much do you want to control the boundaries of your community? Does your community need:	☐ to be private and secure?　☐ open boundaries?　☐ both private and public spaces?	
• How does your community need to interact with other communities? Do you need common tools for sharing and learning with them?		

Technology aspirations

☐ Technology savvy: What are your community's technology interests and skills?

• How interested is your community in technology?	
• What is their capacity for learning new tools?	
• What is the range of skills? If their interests and/or skills are diverse, could it cause conflict or distraction?	

☐ Technology tolerance: What is your community's patience with technology?

• How tolerant are members of the adoption of a wide variety of tools?	
• How many technological boundaries are they willing to cross—for example, sign in to more than one web-based tool, learn to use new tools, or give up old favorites? This helps you understand what level of integration you need.	

Tip: Little things can have big effects. Having an extra login, URL, or tool can discourage participation. Making something a little bit easier can make a big difference.

☐ Technology factors: What constraints are imposed by technology factors?

• What are your members' technology constraints (e.g., bandwidth, operating systems, etc.)?	
• How much time are members able to be online and from where (office, home, field)? Some people have limited online time, or are able to be online only in specific locations. Others are always on. Very diverse situations can affect participation.	

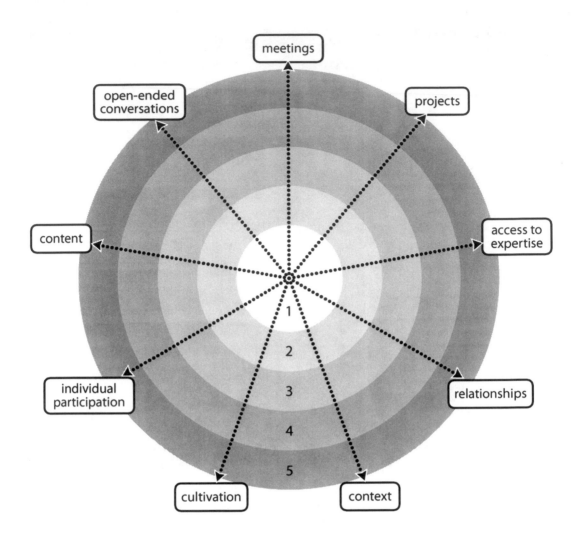

1.2 ➜ Orientations   (Chapter 6)

You can use the following chart and table to create an orientation profile of your community. If you don't recall what the orientations or their variants mean, refer to Chapter
6. First you can use the chart to create a "spider diagram" of your community's profile
by rating each orientation (from 0 being irrelevant to 5 being very important) and then
joining the dots. You can also use the chart to ask different community members to create
spider diagrams of their perceptions of the current state of the community. You can also
ask them to chart their desired states. It is interesting to compare their views and discuss
them. If you want to get more detailed, you can also use the table to check the relevant
variants, and describe the key activities.

0	1	2	3	4	5	Orientations	Variants	Key activities
						Meetings	☐ Face-to-face/blended ☐ Online synchronous ☐ Online asynchronous	
						Open-ended conversation	☐ Single-stream discussions ☐ Multi-topic conversations ☐ Distributed conversation	
						Projects	☐ Practice groups ☐ Project teams ☐ Instruction	
						Content	☐ Library ☐ Structured self-publishing ☐ Open self-publishing ☐ Content integration	
						Access to expertise	☐ Questions and requests ☐ Access to experts ☐ Shared problem solving ☐ Knowledge validation ☐ Apprenticeship/mentoring	
						Relationships	☐ Connecting ☐ Knowing about people ☐ Interacting informally	
						Individual participation	☐ Levels of participation ☐ Personalization ☐ Individual development ☐ Multimembership	
						Community cultivation	☐ Democratic governance ☐ Strong core group ☐ Internal coordination ☐ External facilitation	
						Service context	☐ Organization as context ☐ Cross-organizational ☐ Other related communities ☐ Public mission	

1.3 ➡ Technology configuration: inventory   (Chapter 4)

It is useful to inventory the current technology configuration of your community as a way to understand it better. If yours is a new community, it may not have any specific technology yet, but even for brand new communities, the current configuration may not be empty, for instance if general tools like email or phone are going to be used. You can use a version of the table on the next page to inventory and analyze the current configuration of your community:

1. Get the big picture. Make a list of all the platforms and stand-alone tools in your community's configuration.

2. For each platform, list the tools and check the ones that are being used. Why are some not being used? Are there duplicates? Are there issues around integration between tools?

3. To the left, make a note of which community activities/orientations the tools currently support in your community.

4. To the right, identify the key features of tools. Are some of these features commonly or rarely used? What are the reasons for that?

5. Assess actual tool use. Identify which are dominant and which are only used by smaller groups and individuals.

Tip: It is better to do the technology inventory after you look at orientations so you don't let the technology shape the definition of actual and potential orientations.

Platform 1:			
Supported Activities	**← Tools →**	**Key Features**	**Usage Notes**
	☐ _____ ☐ _____ ☐ _____ ☐ _____		

Platform 2:			
Supported Activities	**← Tools →**	**Key Features**	**Usage Notes**
	☐ _____ ☐ _____ ☐ _____ ☐ _____		

Platform 3:			
Supported Activities	**← Tools →**	**Key Features**	**Usage Notes**
	☐ _____ ☐ _____ ☐ _____ ☐ _____		

Etc...

Stand-alone tools			
Supported Activities	**← Tools →**	**Key Features**	**Usage Notes**
	☐ _____		
	☐ _____		
	☐ _____		

Etc...

1.4 ➡ An emerging picture   (Chapters 4-6)

When you are done with the previous steps, use the following table to compare the Orientations worksheet (1.2) with the current configuration in the Inventory table (1.3):

Covering the orientations	
Compare the left-hand column of the Inventory table (1.3) with the right-hand column of the Orientations table (1.2). What do you notice about the match (or mismatch) between your dominant community orientations and the current configuration of tools?	
• How well does the technology inventory cover the orientations?	
☐ Are you almost there? ☐ Are there big gaps?	
• What is the range of skills? If their interests and/or skills are diverse, could it cause conflict or distraction?	

Achieving integration	
Look at all the pieces of your configuration.	
• What level of integration and interoperability has been achieved?	
• Where are there big gaps?	

Balancing the polarities

• How is the configuration balanced with respect to each polarity?

synchronous ⟷ asynchronous

participation ⟷ reification

group ⟷ individual

• How well does this balance fit your community?	

Note:	This emerging picture becomes the input to Step 2.3.

Now that you understand your community, you are ready to proceed with technology planning.

Chapter 10. Action notebook

Step 2. ➤➤ Provide technology

With a good picture of your community and its aspirations, you can start the process of providing technology.

2.1 ➤➤ Resources and constraints (Chapters 3 and 7)

First consider the resources and constraints in your environment that will influence your decisions.

Organizational context		
☐ Within an organization:		
• Do you need to develop your technology strategy in collaboration with the IT department?	☐ High level of control ☐ Some flexibility ☐ Relative freedom	
• What specific resources and constraints come from the IT department? (Get these details in writing if you can).		
• What community-oriented technology do they have already? Is it usable?		
• Does community-oriented technology need to interoperate with other enterprise software?		
• What standards do you need to adhere to?	☐ firewalls and security standards? ☐ databases or data standards? ☐ single login protocols? ☐ company look and feel? ☐ policies?	
☐ Across organizations:		
• Which organizations can host the community or provide resources?		
• What strings are attached?		
• What problems can boundaries create across organizations:	☐ firewalls and security? ☐ data standards? ☐ login protocols?	
☐ Outside any organization:		
• Where will resources for technology and for tech stewardship come from?		
• What tools can members contribute and what will happen if they leave?		
• What open web standards do you need to adhere to?		
• Do you want to "brand" your community through its look and feel?		
• If you are an open community, how will you deal with spam?		

Financial strategy	
☐ **Investment factors: What are your financial constraints and plans?**	
• What is your budget?	
• What are your short- and long-term goals and investment strategies?	
• What are the must-haves for today, and what are longer-term needs that could be deferred?	
☐ **Installation factors: Are you planning to acquire software?**	
• Who will install and configure your software?	
• Where will the software be hosted? ☐ Hosted service (ASP)? ☐ On your own servers?	
• What is your plan for ongoing technical support?	

2.2 ➻ Select an acquisition strategy   (Chapter 8)

Taking into consideration both your circumstances and the options available in the marketplace, shape an acquisition strategy by selecting one or more from the list below. If you select more than one to create a composite strategy, you will need to think about integrating the outcomes.

• Do you want to get up and running quickly but aren't ready to invest in technology yet? ➡	☐ **Strategy 1: Use what you have**
• What are members already using in their daily lives (email and telephone)?	
• What might hosting organizations let you use?	
• Could you repurpose an existing tool or make small adjustments for your community's use?	
• Do skill gaps in the community prevent an existing tool from serving the community fully?	
Tip:	If this is not enough, combine with strategy 2 (using free platforms) and/or 7 (patching pieces together).

• Do you need something that works across organizations and requires no money? ➤	☐ **Strategy 2:** **Use free platforms**
• Are these tools widely accessible enough?	
• Can you live with some advertisements?	
• How important is control of your community data?	

Tip: Make sure to pick options that allow you to back up your membership list, archives, and so on.

• How much work is it to use/support these tools?	

Tip: Look for tools with active user communities. They usually can offer quicker support than the company providing free tools to such large numbers of people.

• Does your community live in an organization with an existing IT infrastructure? ➤	☐ **Strategy 3:** **Build on an enterprise platform**
• What parts of the infrastructure could you reconfigure to suit your community?	
• Have you built relationships with people in the IT department and sought their support?	

Tip: IT folks are a key to this strategy, and it is worth investing time in those relationships.

• Are other communities in your organization using the enterprise platform?	

Tip: If you want to do additional development or customization, it might be easier if you pool resources.

• Do you want one platform with a variety of tools and features all bundled together? Is that convenience critical to your community? ➡	☐ **Strategy 4:** **Deploy a community platform**
• Is the platform as good as it looks?	
• Is the functionality what you need?	

Tip: As you shop for a community platform, make sure you can try it out. Look carefully not only at the tools offered, but the features that make them usable.

• Does the platform and the vendor have a history of focus on your critical orientations?	

Tip: Get feedback from other communities that have used the platform, especially communities that show similarities in their orientations.

• Do you have very unique needs that are not met by tools in the marketplace? • Do you have deep technological knowledge in your community or access to financial and technical resources? ➡	☐ **Strategy 5:** **Build your own**
• Are you sure you are ready for this? Really sure?	

Tip: Define your needs first in terms of your community orientations and activities, then in terms of technological functionality. Work closely with your developers throughout the process so they have clarity on the tools' functionality you're looking for.

• What are your long-term plans to support a custom-designed platform?	

Tip: Make sure more than one person knows the specifics of the system, so you are not stuck when a key member or a developer leaves the community.

Chapter 10. Action notebook

• Does your community wish to benefit and contribute to a larger network of people using the same software? ➤ • Do you have a philosophical preference for free or open-source software?	☐ **Strategy 6:** **Use open-source software**
• Do you have the technical skills required to customize current open-source offerings?	
• Have you allocated some of your time to being involved with the open-source community?	

Tip: Participating in the open-source community is the best way to use the software.

• Are you interested in new tools that quickly allow you to combine new functionality into basic tools like blogs and web pages? ➤ • Do you like quick, low-cost experiments?	☐ **Strategy 7:** **Patch pieces together**
• How will you test the functioning and usefulness of a new tool that you patch into the existing mix?	
• Who will do the addition of pieces and how will that be negotiated?	
• How do you balance potential benefits/cost to the community of dealing with new things or things that just "sort of work?"	

Tip: Balance innovation with the community's attention and energy.

2.3 ➤ Seek a solution

It is now time to apply your chosen strategy and all the information you have gathered so far to put together a technology configuration for your community.

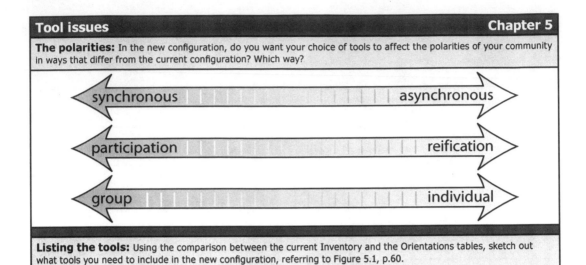

Tool issues   Chapter 5

The polarities: In the new configuration, do you want your choice of tools to affect the polarities of your community in ways that differ from the current configuration? Which way?

synchronous   asynchronous

participation   reification

group   individual

Listing the tools: Using the comparison between the current Inventory and the Orientations tables, sketch out what tools you need to include in the new configuration, referring to Figure 5.1, p.60.

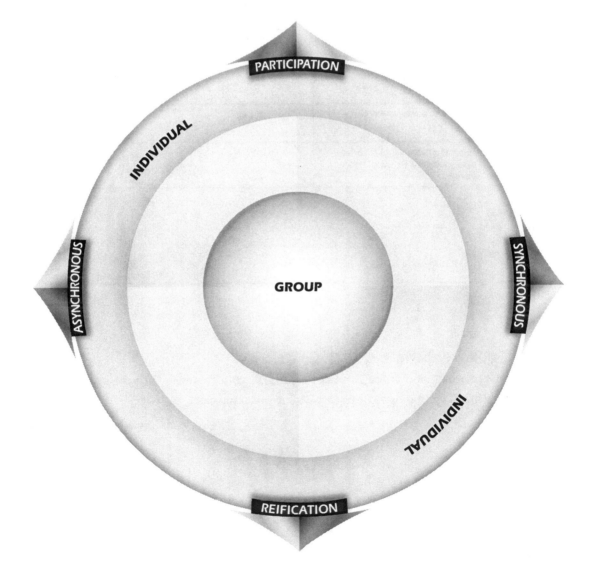

Tip: This is an iterative process where assumptions of linearity are misleading. Looking for tools is likely to reshape the picture of your community: the relative importance of orientations, the desired balance of polarities. You may even discover new orientations or imagine new uses for existing tools. The interplay between technology and community evolution is truly a mutual process.

Platform issues	Chapter 4
☐ **Adequacy:**	
• Is there one platform that has most of the tools that your community needs?	
• Using the features questions below, are the versions of these tools adequate?	
• Which critical tools are not covered?	
• Which are extra (there, but currently not needed)?	
☐ **Integration:**	
• How simple (or intuitive) is the platform to use?	
• How well does it combine the tools that your community needs?	
• Are key features such as menus, navigation cues, new material indicators, graphic elements, and controls deployed consistently and appropriately across the platform?	
• Can tools be turned on or off at will?	
☐ **Performance:**	
• How many concurrent members can it handle? How much activity?	
• Does it support multiple communities and are new ones easy to launch?	
☐ **Access:**	
• Can subcommunities be formed easily?	
• Can individual access rights be assigned flexibly to various spaces and items?	

☐ **Pricing:**		
• Is the pricing structure per:	☐ free ☐ community ☐ seat ☐ activity ☐ platform ☐ other	• What are the implications for your community?
• What is included in the price:	☐ hosting ☐ support ☐ upgrades ☐ other	• What other costs are not included?
• How does the overall cost compare with alternative platforms?		

☐ **Vendor relationship:**	
• What is the reputation of the vendor?	
• What ongoing relationship do they offer?	
• Are they willing to develop the platform and work with you as your needs and technology evolve?	
• Is the data in a standard format that can easily be moved to another platform/vendor?	

Features issues	**Chapter 4**
☐ **For any given feature:**	
• Does the feature support the specific ways in which a community conducts its activities?	
• Does it add or reduce complexity? Can it be turned on or off to make the technology more useful or easy to use?	
• Does a feature inherently appeal to beginners or to more experienced users? What members' skill levels make a feature valuable?	
• Is the absence of a feature problematic for the usefulness of a tool or platform?	
• Will members expect a feature or recognize it because of previous experience?	

☐ **Completeness:**	
• Are some important functions missing in the overall configuration?	
• Do some tools duplicate each other, and if so, could subgroups evolve from using different tools for the same purpose?	
☐ **Integration:**	
• What level of integration is required between existing tools and platforms in the configuration?	
• Where are there integration gaps and how are you going to address them?	
• How compatible is the configuration with other platforms or tools that members use?	
• Do features that support integration across tools or platforms have the quality and consistency that you need? Do security features, for example, conflict with tools such as RSS feeds?	
• Are there features that help make content portable across tools (e.g., content from conversation board to wiki)?	
• Can members import content from other tools into the configuration?	
☐ **Security:**	
• Is the overall configuration secure enough for your purpose?	
• Are some security features likely to get in the way of the community's togetherness?	

2.4 ➡ Plan for change   (Chapter 9)

Major technology changes are about more than technology. They involve change management.

☐ **Timing:** Time your transition/implementation to fit your community schedule		
• Are you ready for the attention you will get when technology issues move to the foreground during a major transition?		
• Are there times when "messing with technology" will ...	☐ be a good community-building activity? ☐ cause a problem?	
• What external events or schedules do you need to take into consideration (budget cycles, holidays, availability of support, for example)?		

☐ **Implementation:** Plan for the change process	
• What are your plans for the practical implementation of the new technology?	
• Who are your main partners for the implementation process?	
• Do you know enough about your community to know what to expect?	

Tip: Talk to as many people as you can.

• If you have to make a lot of assumptions, how are you going to leave room to adjust as you move forward?	
☐ **Learning:** Plan for a learning curve	
• Will new tools affect their work and community focus in a significant way?	
• How much beta testing can you do or do you want to do? Can you test software from a vendor or in other communities using it?	
• How will you orient, train, and share good practices with your community?	
☐ **Integration across tools:** Help the community develop new practices	
• Are there integration issues in the new configuration?	
• How do you plan to help the community develop new practices to achieve a productive level of integration?	

Now that you have provided technology for your community, you need to continue with stewarding technology in use.

Step 3. ➡ Stewarding technology in use

3.1 ➡ Everyday stewardship (Chapter 9)

Even when technology is not the focus, the work of stewardship goes on in the background. A number of tasks require attention.

☐ **New members:** Support new members in their use of the community's technology	
• How many new members do you have per month?	
• Does the community have a welcoming activity for them?	
• How do you plan to onboard them on the community's technologies?	
• What is the minimum they need to know to be able to participate meaningfully?	
• What resources do you have for this? Who can help you?	

☐ **Practice:** Identify and spread good technology practices	
• How are you going to identify the new practices that the community is developing to use technology, especially ones that might be going unnoticed?	
• How are you going to share and spread them unobtrusively?	

☐ **Experimentation:** Support community experimentation	
• Is your community changing? Is it curious about new tools?	
• How will you support technology experimentation without disrupting the whole community?	

☐ **Boundaries and access:** Attend to community boundaries created by technology	
• How will you manage access as the community and people's roles evolve?	
• What unexpected boundaries does technology create?	
• Do technology preferences or skills create boundaries?	

☐ **Technology integrity:** Assure continuity across technology disruptions	
• Who has administrative permissions so they can help you "keep the lights on" over time?	
• How do you make sure vendors get paid on time and domain registrations don't lapse?	
• What are your practices for system backup?	

3.2 ➡ Community end-of-life closure

Community end-of-life situations have important implications for tech stewardship. Work with the community to design a process for "turning off the lights."

☐ **Shutdown:** Attend to disposition of the community's technology resources	
• Is the community ending or merely going dormant? Who can decide?	
• When should online spaces be closed down?	
• Who will cancel contracts with technology services such as ASPs?	
• What to do about member profiles and account information?	
☐ **Community history:** Pay particular attention to the preservation of community artifacts	
• Does the community want to archive and preserve parts of its history?	
• How will you identify which artifacts should be preserved?	
• Do privacy agreements or concerns dictate that you erase sensitive materials?	
• Where and how will the material be archived? Should members receive a copy of the archive?	
• Who will have access to the archive?	
• Who will take care of it? Could other communities have an interest in being the caretakers of a legacy left by a defunct community?	

Good Journey!

Part IV:
Future

A more distributed future

In the final two chapters of the book, we step back from the day-to-day prac-ticalities of technology stewarding and take a more conceptual look at the trends in the search for new digital habitats at the intersection of community and tech-nology. The chapter extends the history started in Chapter 2 by exploring how these trends interact with the polarities introduced in Chapter 5—and how this might change our understanding of both community and technology.

We started the book by reviewing the history of mutual influence between community and technology. We suggested that much technology innovation and development has been stimulated by the needs of communities. In turn, each technological development and increment of change has affected community life. We illustrated this process with some of the stories about how early Internet and World Wide Web founders were innovating in the service of their communities. We also sprinkled the book with small vignettes of how individuals and their communities have been both inspired by and inventive with technology.

A vortex of inventiveness

Viewed as a trend, the history of mutual influence between technology and community creates a vortex of inventiveness that propels both forward. The inventiveness of the technology sector is changing the possibilities for communities, what they look like, and the practices that allow them to function. In turn, the inventiveness of these new, increasingly technology-mediated communities is influencing the path of development and use of many technologies. As these two sources of inventiveness converge, their momentum creates an accelerating vortex that disrupts and reshapes everything in its path.

Deliberate technology stewardship plays a key role in this collective inventiveness. This happens on multiple levels at once. It happens within communities where tech stewards articulate the needs of their communities and project them on the landscape of available technologies. It happens in the marketplace where tool builders and open-source communities respond to and generate needs with new offerings and new ideas. It happens competitively among vendors and collaboratively in standards bodies that attempt to develop compatible interfaces and interoperability protocols. It happens in organizations exploring integrated platforms and infrastructures. It happens in interactions between tech stewards and vendors and in conversations among tech stewards. Even a community that resists or rejects an innovation offered in the marketplace is participating in the overall process of inventiveness, perhaps by winnowing out ineffective technologies. This aggregate inventiveness reaches far beyond any given community, tool, designer, organization, or practice. It represents tremendous collective energy that is gathering momentum.

The roles are blurring

The momentum that we are seeing involves many different actors and a complex mix of roles. A remarkable trend in this process is that the players' roles are blurring. Increasingly, the same people participate in building, using, and refining technology. Indeed, contributions are no longer restricted to people with traditional programming skills. Tech stewards can participate in the design of tools; users can build tools and make them available. Tool-builders participate in communities that use their tools; they often use their own tools to collaborate in their tool-building activities, experiencing and understanding the user perspective in the process. While people who are actively and intentionally involved in technology stewardship for communities still have a special role to play, such stewardship is becoming a more diffuse responsibility at the same time as its importance is increasing.

The process is accelerating

As the roles of users, tech stewards, and tool builders are blurring, and engagement among them is increasing, feedback loops are becoming shorter: the momentum of inventiveness is accelerating. We see this in the exploding number of offerings in the marketplace and the mix of excitement and confusion it causes. We also see it in the increasing attention that community leaders and members pay to technology and in the sense of overwhelm that some of them experience when faced with so many tools and options.

This acceleration both reflects and intensifies the convergence between Internet technology and the peer-to-peer nature of learning in communities of practice. Why this convergence is intensifying today is an interesting question, with a range of possible factors. It could be due to a combination of:

- The role of some communities in influencing the direction of technology development

- A growing awareness of the social nature of web technology among tool builders and Internet companies becoming increasingly attuned to it

- Increased attention to the practices necessary to take advantage of the new medium, what it affords, and its tools

- Widely spreading web literacy in society at large

- Changing expectations and beliefs about what can be accomplished at a distance

- The sheer variety and social nature of the new tools being developed

This intensifying convergence makes for a very creative time. It presents both opportunities and challenges for communities: the potential to expand the experience of community beyond our imagination and the risk of tearing things apart when the rate of change exceeds the communities' capacity to adapt.

Trends in digital habitats: reconfiguring community polarities

Reflecting on the history of technology and community, we see several interacting trends that both arise out of this history and give it momentum going forward. These trends and their components are listed in Table 11.1, which organizes them into four major areas. Note that the first three areas relate to the three community polarities introduced in Chapter 5:

- Togetherness and separation in time and space

- Participation and reification

- Individual and group identities

The fourth area—the nature of a socially active medium—relates to all three polarities. Because these polarities are fundamental community concerns that make technology relevant, we see them as key drivers in the inventiveness that ties community and technology together. As they inspire the development of social technologies, these three polarities are stretched and reconfigured.

The next four sections investigate the trends in each of the four areas defined in the table. After a brief explanation of each of these trends, we focus on their interplay with the polarities.

1. Increased connectivity across time and space	
• Ubiquitous connectivity	From intermittent connections using modems to "always on" access through wireless and mobile technologies
• Virtual presence	From purely text-based interaction to virtual presence, multimedia experiences, and avatar-based environments
2. New modes of engagement	
• Generalized self-expression	Easy publication to the whole world: the spreading blogosphere, and "personal space" sites
• Mass collaboration	Wikis, tagging, social networking sites, publicly shared, interactive storage spaces
• Creative reappropriation	Remixes, social bookmarking and personalized lists, mashups
3. Changing geographies of community and identity	
• Homesteading of the web	Proliferation of sites, tools, and links; multiplicity of places for any topic; emergent patterns of meaning and interrelatedness
• Dynamic boundaries	Boundaries defined by activities and their traces, including the tools that rank locations and direct traffic
• Individualization of access	RSS, personalized aggregation, customized search, personalized access to sites
4. Toward a socially active medium	
• Social computing	Social relations and interactions as data, "folksonomies" and tagging, networking services, distributed decision processes, reputation computing, socially directed search
• Semantic web	Meaning-based representation, intelligent agents, new-generation search
• Digital footprint	Trails of our web activities that become an expression of our identity online

Table 11.1. Trends affecting the polarities of communities

1. Increased connectivity across time and space

Two main trends are changing the character of community habitats in time and space: new forms of ubiquitous connectivity and new forms of virtual presence.

Ubiquitous connectivity. After the early bursts of expensive modem-based connectivity, the current trend is toward "always on," whether this ubiquitous connectivity is provided by fixed devices in our offices and at home or is carried with us through mobile devices with wireless access. A multiplicity of forms of togetherness is constantly available. We can access documents or add our two cents to a conversation any time, from any place. We can share with the world "what's up" in 140 characters on Twitter, or just share with our select networks. Our communities can always be with us. Whereas our experience of an open-ended conversation orientation was place-based and intermittent (either face to face or in a discussion board), now it can be continuous and uninterrupted. This brings opportunities such as rapid sharing and cross-pollination of ideas across diverse individuals and groups. It presents challenges dealing with the unprecedented communication volume made possible by ubiquitous connectivity and the number of communities and networks a person can belong to simultaneously.

Virtual presence. Technology is offering an ever-growing variety of ways to be in each other's presence across space and time. Virtual co-presence started with early chat rooms involving typing text on computers and transmitting it via modems. It evolved into multimedia web-conferencing software that uses a mixture of audio, video, text, and shared applications to offer new ways to experience meeting as co-practitioners online. We have hardly started to explore the potential of avatar-based virtual environments such as Second Life. Seeing others' avatars, even if we do not interact with them, lets us know we are not alone. No one knows where this trend will lead, but it is clear that it has the potential to transform the way we interact, and more generally, the way we experience togetherness.

Polarity: Dynamic fluidity of togetherness and separation

These trends contribute a dynamic fluidity to our experience of togetherness and separation. The very meaning of togetherness and separation is changing, and the distinction between the two is less clear. We are more easily together and more easily separated, as when a conversation is interrupted by a ringing phone. Sometimes separation and togetherness are confounded, for instance, when we multitask while participating in a web meeting, or when we blog while sitting in a conference.

Our experience of space is becoming a dynamic mixture of physical and virtual relationships, of synchronous and asynchronous connections, where our togetherness is fleeting at the same time as it is becoming "always on." A range of different experiences of space and time becomes an integral part of our digital habitats. A teleconference becomes a podcast. New tools enable us to co-edit a document on the web in a fluid sequence of synchronous sessions and asynchronous work. We move rapidly between visibility and invisibility in a blending of the physical and the virtual.

As a consequence, participation is increasingly spanning geographic, organizational, professional, cultural, and national boundaries. If new technologies and practices redefine what it can mean to learn together, a key question is how attentive we are to our togetherness, particularly as we multitask. What are the implications for the value of our interactions? Is the benefit of more complex participation diminished by reduced attention? Can the fleetingness and multiplicity of togetherness become a source of separation?

2. New modes of engagement for interacting and publishing

The evolution of technology has increasingly put new abilities to publish, participate, and interact into the hands of anyone with a computer connected to the Internet. Today we no longer need to rely on specialized professionals to put up a web page. Community activists connect their volunteer cohorts with blogs and email lists. Global activists coordinate nearly instantaneously and respond to emerging issues with the click of a mouse. Scientists share data in real time through wikis. Artists create, disseminate, and even sell their work to global audiences, and invite comments and participation in the creative process. We see three aspects of this trend affecting communities and learning in social contexts.

Generalized self-expression. The web is becoming a medium of self-expression with an unprecedented reach. The blogosphere is expanding very rapidly, creating a multiplicity of individualized voices throughout the web. Sites that offer people the ability to publish multimedia material with social affordances for cross-linking and commenting are part of the same trend, whether they enable the creation of personal web pages, the sharing of photographs, or the downloading of videos. This creates the possibility of broad self-publication and self-expression in the context of constant interactions that mix the social and the individual.

Mass collaboration. At the same time as the web enables individualized self-expression, technology enables collaboration on an increasing scale that produces a proliferation of communal creations. By "mass collaboration," we refer to the number of potential collaborators one can find for working on a project of any size. The ability to broadcast a problem and ask for help is characteristic of this aspect of mass collaboration. Technology is also enabling collaboration at an unprecedented scale for a given project.[1] Wikis are a quintessential example of this phenomenon, but the broadly collaborative development of extensive shared tagging and social bookmarking catalogs is representative of the same trend. How do our collaboration practices reflect the growth in the tools and volume of collaboration opportunities? Does collaboration on a massive scale demand a new sense of identity where authorship is primarily communal?

Creative reappropriation. The digitization, availability, and linkability of web resources, including the use of widely accepted standards such URLs as linking devices and application programming interfaces allow communities and their members to reuse and remix web content, for instance through social bookmarking, personalized lists,

---

1. For an extended exploration of this trend, see the book by Don Tapscott and Anthony Williams. Wikinomics: how mass collaboration changes everything (New York: Penguin Publishers. 2006).

aggregated feeds, and even mashups that combine applications into new services. Therefore, the resource base of a community is not a centralized library that it owns but a community-specific view of the web, driven by relevance to its domain and collaboratively organized by its members as expressed through social bookmarks. Often, these are publicly viewable resources whereas in the past they might have been placed in private spaces.

Polarity: Reweaving participation and reification

The new modes of engagement are reweaving the distinction between participation and reification, sometimes with rapid cycles of interaction and publication, sometimes with modalities that combine both, such as with wikis. Back in the days of letter writing, publishing and communicating were also part of the same process, but letters were slow, usually written to one person, and comments meant another letter in return. Now much of our communication can be both collective and instantaneous.

Published self-expression becomes a means of interactive connectedness as we comment on each other's public diaries in distributed conversations that span across blogs. We interact as we edit each other's contributions to a wiki-based document or through our participation in a distributed tagging process. We make statements by reusing, pointing to, and remixing resources. Mutual engagement in a community entails an intertwined combination of resources and interactions—exchanged at the same speed through a medium where the distinction between them is not inherent but more and more a matter of interpretation.

3. Changing geographies of community and identity

Three complementary trends are shaping the web as an evolving habitat, where communities and individuals shape their identities in relation to each other.

Homesteading of the web. The term "homesteading" refers to the process by which settlers would claim a piece of territory under the US government's Homestead Act of 1862. Similarly, the web has become an ever expanding frontier, opening an expanding territory of topics to grab and claim ownership of. Indeed, we are witnessing a rapidly expanding proliferation of sites, pages, tools, groups, and links that are progressively populating the web on every imaginable topic. Sites compete to be the authoritative place for the latest hot topic. This increasing complexity produces emergent patterns of content, interactions, and interrelatedness that give rise to a multiplicity of places of focus, for any one domain. People can connect in multiple ways and around multiple locations. This plethora of resources and places yields richness, overabundance, and in some cases fragmentation for communities and individuals.

Dynamic boundaries. Popularity shapes visibility and accessibility, since traffic, linkages, and attention (or lack of it) are sources of dynamic boundaries produced by new tools such as search engines, ranking algorithms, and activity reports. Popularity creates boundaries by excluding less popular voices, but it also enables new voices to break through rapidly, provided they can garner enough traffic.

Individualization of access. While the complexity of the web is expanding, tools that enable individuals to navigate this complexity in personalized ways are also proliferating. Access to information gets channeled through subscription mechanisms such as RSS and recombined through aggregators that create a personalized vista on the web. Search and access to sites are also increasingly customizable.

Polarity: Dynamic group formation and multimembership

The new geographies of community and identity enable both a stronger focus on the individual who can work across the resulting networks, and new possibilities for the groups that can form within them. This phenomenon produces more individualized trajectories of participation at the same time as it enables easier group formation, more group interactions, and belonging to multiple groups and communities in multiple ways.

Technology has afforded wider gradients of participation in communities and therefore new possibilities for orienting to individualized participation. This includes personal portals and aggregation tools and the ability to personalize one's experience using a tool designed for a group. It has made new forms of peripherality possible through what is known as lurking (or even merely surfing) as well as quick moves from peripherality to active participation and back.

The combination of homesteading, individualization of access, and dynamic boundaries makes for a landscape that is constantly shifting. The "shape" of a domain is reconfigured as new locations and individuals engage with it—a blog here, a series of comments there, a new site somewhere else. This makes it easier to participate peripherally—to grab bits and pieces here and there—but it is also more difficult to see the whole shape.

The individual consequence of these developments is that the potential for multimembership has grown as we are able to "belong to" more communities than we can really manage—communities that are themselves in flux. Again, more possibilities may lead to less attention as we try to deal with increasing volume. Whether disrupting or empowering, this dynamic multimembership creates a new geography of identity that is both quite social and at the same

Chapter 11. A more distributed future

time very individualized. Indeed, the nexus of multimembership of each individual becomes a unique intersection—an effort, more or less successful, to make sense of the multiplicity and create threads of coherence.

4. Toward a socially active medium

As people increasingly use the web to connect with and find each other, to express themselves interactively, and to form communities, it is becoming a medium that shapes the social world, by usage and by design. The combination of distributed production, digital representation, and search capability makes the web an active medium where the social and the informational build on each other.

Social computing. Social software is increasingly transforming the web into a social medium—social relationships become part of the medium and its computational capabilities. Programs produce and use data about social relations and interactions. For instance, socially directed search uses such data to improve the likelihood of more targeted search results. Communities can build folksonomies through shared tagging and evaluation of content. They can make decisions through distributed decision processes such as polling and prediction markets. Networking services and reputation computing can guide the seeking and building of relationships. Lists of friends become part of a "social calculus." All this blurs, for better or for worse, relationships, information, personal spaces, and identity into a computable resource.

Semantic web. This socially active nature is reinforced by recent efforts to introduce semantic computation into web standards. For example, a project is developing stan-

Venturing into new community geographies

Helen Walmsley is the leader of a community aimed at engaging educators at Staffordshire University in e-learning best practices. By welcoming people from around the world into the community, university educators are connected with authentic practice in all its variety. The Best Practice community hosts a variety of synchronous and asynchronous activities on the community's Moodle platform. As an experiment and a demonstration of potentially useful technologies, Helen recently opened groups for the community on Facebook and Ning, encouraging community members to explore those new platforms. The new platforms partly duplicated existing functionality but also extended it; for example, Facebook provided a better mechanism for announcing events. But the big surprise was how the community grew as a result. Members who had drifted away suddenly re-appeared and new people joined, bringing new energy, ideas, and perspectives into the community. An important characteristic of people that connect through Ning or Facebook is that they are well-connected with other active communities on those social networking sites. There is no requirement that everyone register on all three platforms. Indeed, many members are registered on only one, so they have only a partial view of the full community. Members show up for a community event due to an announcement or activity on only one of the platforms. In this case, the three platforms do not contribute different functionality. Rather they are three "beachheads" into different social worlds defined by distinct social-networking platforms. The partial overlap among these distinct community beachheads makes for a rather complex membership structure.

http://crusldi1.staffs.ac.uk/bestpractic-emodels

http://ning.com

dards to describe "friend of a friend" relationships among people, through profiles that use standardized categories.[2] Another provides a vocabulary to describe "semantically inter-linked online communities" in terms of the web spaces enabled by social software, such as discussion forums, blogs, RSS, social bookmarks, and image galleries.[3] This will push the active interplay of the social and the informational even further by introducing some social intelligence into programs and search capabilities. We have hardly begun to explore the implications for communities of such a socially active medium.

Digital footprint. Our engagement with the socially active medium created by new technologies leaves traces each time we do something on the web. These traces become an impressionistic picture of the self–one that is scattered like dots of paint in a networked canvas, which includes discussions, product reviews, blog posts, pictures, podcasts and videos, instant messages, and tags, as well as comments from other people posted on our traces and the comments we add to others' traces. This "digital footprint" is an evolving (but enduring) image of ourselves over which we have only very partial control. Admittedly, we have always participated in many conversations and interactions; we have always had multiple means to store our memories; our identities have always combined what we produce ourselves and what others reflect and project on us. Recorded in a socially active medium, however, our traces are searchable; they can be found and reassembled dynamically; they are inspectable, manipulable, and remixable. Even when we think we have deleted them, they are found again. Scattered and computable, our footprints create reconstructable trajectories in a public space, largely out of our control. Who are we in this mirror that remembers and talks back with a voice that is only partly our own? Does the potential to remember so much mean that we know ourselves and each other better? Or could our digital footprints hide as much as they reveal, as if their very transparency only added to the mystery of identity?

Polarity: Ability to find each other and to see the social fabric

Social software such as networking tools, profiles, search engines, and track-backs makes it possible for people to find each other, discover new connections and commonalities, and explore the value of forming communities. Even tagging or a remix of RSS feeds can be a convening force by revealing a common interest and its potential as the domain for a community. Related communities can now hook up with each other and create new learning agendas together.

The increasing ability to find people and things means that it may come to matter less where you interact and where you publish than how findable and

---

2. See the project website at http://www.foaf-project.org or its description on Wikipedia at http://en.wikipedia.org/wiki/FOAF_%28software%29
3. See the project website at http://sioc-project.org or its description on Wikipedia at http://en.wikipedia.org/wiki/SIOC

linkable your content and interactions are. Publishing and interacting are both social actions that become part of our digital footprints.

These traces embedded in a socially and computationally active medium produce new representations to visualize networks, individual participation, and community history. These "meta-representations" apply to both groups and individuals: whether or not they are desired, they provide new ways to see the geography of community and identity. They become tools for community development, for navigating the landscape, and for fostering new connections. They also take on a life of their own— shadows with memory and computational capabilities that give them an independent reality.

Challenges and opportunities

Inventiveness within the concurrent evolution of community and technology produces a momentum that can wreak as much havoc as it can create new possibilities for community cohesiveness. The outcome is not inherent in the technology alone but depends on the practices that communities develop to take advantage of new technological developments. Some of the key challenges include:

- **Overwhelming volume.** While new technologies theoretically allow us to share the load of filtering, absorbing, and making sense of all that flows across our computer screens, even the choice of how to use those new tools is complex. Some of us have difficulty letting go of the need to control the flow of information or to "see everything." Particularly when we belong to multiple communities, we can fall into the trap of spending so much time absorbing information that we have less time practicing and interacting, whether online or offline.

- **Falling into groupthink.** New tools that allow us to deal with volume by focusing more intentionally enable us to view just the slice of the world that matches our expectations, perhaps blinding us to other views. Online gatherings of large groups of people who are interested in the same subject can create an illusion that the group is "the whole world."

- **Vulnerable systems.** Connected systems are vulnerable to disruptions. Both technological attacks to the Internet's infrastructure and the applications it hosts, as well as attacks on groups by either members or outsiders, can disrupt or even destroy online groups. While networks are designed to withstand many of these attacks by routing around the point of attack, communities are much more vulnerable.

- **Stretching our relationship limits.** The number of potentially useful and meaningful relationships becomes so large that being selective becomes more important, even as we increasingly rely on those very problematic sources of information to select partners. Connections in passing with so many other people can swamp meaningful interactions with a few.

A number of trends affect the interplay of communities and technology in four areas: connectivity, modes of engagement, geographies of community and identity, and the social nature of the web medium. We need to follow these trends, notice new developments, and understand the positive and negative effects of the momentum they create. It is an aspect of good technology stewardship to understand this vortex of change and respond proactively—to seize its opportunities and to confront its challenges in the service of our communities. As a way to build a better understanding of this momentum of inventiveness and its implications, we propose a learning agenda for technology stewardship in the next chapter.

A learning agenda

*T*his chapter again is oriented to the future, not to gaze into a crystal ball, but to develop a learning agenda for technology stewardship. Our learning agenda is an invitation to explore three areas where technology stewardship will matter: serving existing communities, making new communities possible, and stretching our very notion of community.

When we reflect on the trends outlined in Chapter 11 from the perspective of technology stewardship, we see a learning agenda emerging. By learning agenda, we mean a collective sense of what we need to learn in order to move our practice forward. Articulating a learning agenda is a useful community-building activity. In the case of technology stewardship, such an agenda can guide our learning as we make the future, both individually and collectively. It also provides a language for tech stewards to talk to each other and to connect with other communities struggling with similar concerns.

A literacy of technology stewardship

Community technologies require new practices for their successful adoption, productive use, and further development. As technology has become inextricably intertwined in the lives of so many communities, we see the need for quantum leaps in practice. We are enthusiastic about the interplay between technology and community because of its inherent inventiveness and the potential it carries. Yet, we also harbor a good amount of skepticism about technology adopted for its own sake and we are aware of potential downsides. The introduction and adoption of all this gear is exciting. However, in the end we are most interested in the invention of new practices, a process that we already see happening in many communities.

To play their key role in the development of these new practices, tech stewards require a new kind of literacy—a flexible understanding about how digital habitats can serve the learning of communities. We have proposed some foundations for this literacy when we introduced concepts like technology configuration (Chapter 5), community polarities (Chapter 6), and community orientations (Chapter 7).

Technology configuration is a simple concept whose purpose is to elevate the conversation about technology to include all of the tools that support the functioning of a community. We have suggested that the full configuration is the proper domain of the tech steward. While the concept is simple, its implications for our learning agenda include complex issues of completeness, dispersion, integration, fragmentation, and boundaries. When can we say that a set of tools adequately covers the activities of a community? How do we deal with the fact that tools are increasingly distributed across the web and in the hands of individual members? How can we make these tools work together seamlessly enough? How should we manage the internal and external boundaries created by a configuration? These questions keep the focus of the tech steward on sustaining an experience of community across the tools members have adopted.

Community polarities provide a framework for assessing the effects and opportunities of technology and associated practices. They suggest pairs of related areas to which we should pay particular attention, either when technology leads to an imbalance in a polarity or when technology blurs or recombines the polar elements.

- When does technology connect a community to the point of overwhelming some of its members or marginalizing others? How can it enable them to feel together while being apart?

- In which cases does technology encourage a community to place too much emphasis on documents and information management? How can it enable more productive interactions around published resources?

- When is the cohesion of the group threatened? When does the group identity become so strong that it prevents openness and learning? When should the needs of individuals be asserted? How can a tool allow group and individual identities to build on each other?

If, as depicted in Chapter 11, technology stretches, reconfigures, and reweaves these community polarities, we will have to learn to modulate them more intentionally in the practice of our communities. For example, we must become more skillful at anticipating the effects of a new tool on a community's dynamics. Like good cooks who know the effects of various ingredients on flavor—sweet and sour, salty and spicy—good tech stewards understand the characteristics of tools in terms of community dynamics expressed by the polarities. This enables them to diagnose the effects of tools and combinations of tools into configurations that serve communities along all the relevant dimensions.

Community orientations similarly provide a perspective on technology that is focused on learning within a community. Does a community need more meetings or does it need support for personal relationships? Could technology better connect a community to its context or make information available for reflection on cultivation processes? Can technology be used to shift an orientation? If so, how? When would that be appropriate? What if an emerging orientation is in conflict with a community's existing technologies? Are there new orientations that emerging technologies and related practices make possible? Again, orientations provide a framework to understand the interplay between technology and community processes, and to leverage this interplay in the service of community development.

It is in this sense that concepts like configuration, polarities, and orientations constitute the foundation of literacy for technology stewardship. Developing and applying this literacy lies at the core of our proposed learning agenda.

The rapidly evolving context created by the interplay of communities and technology described in Chapter 2 and the trends of Chapter 11 further frame this learning agenda in three areas:

1. By offering new opportunities and challenges for existing communities

2. By enabling the formation of new communities we could never have imagined before

3. By stretching our very notion of communities of practice and the learning potential that exists within and among communities

In the next three sections we explore several learning challenges in each of these three areas. Revisiting the trends of Chapter 11, we express this learning agenda in the form of open questions that need attention. Answers to specific questions are likely to vary across contexts and over time, but the perspective as a whole is a useful framework for forging ahead.

1. Serving existing communities

The first challenge is to learn how to serve existing communities better as they navigate through the maze of potentially useful technology. Looking forward, we see more choices and configurations of technologies that are more complex because they combine tools from multiple platforms. We see communities opportunistically hopping from one technology to another to find the tools and features that members need. But since communities can be overwhelmed by too many choices, how do we enable them to take advantage of what makes sense without distracting them from their purpose? Communities of practice have too much work to do to afford much faddishness. If some members are overeager to jump into every new possibility, how do we temper their enthusiasm when it threatens the cohesion of the community? Learning in communities requires too much cohesion to split over uneven adoption, access, or personal preferences with regard to technology.

Technology stewardship requires a balancing act between conservative stability, where communities stick to what they are comfortable with (even if they have outgrown it), and runaway adoption, where members become enamored of technology for its own sake. How do we strike that balance? When should we advocate for or resist change and when just witness from the sidelines? How can we engage a broader group in decision-making as roles are blurring and technology stewardship is increasingly something that members may be able to do? These questions may not have universal answers, always having to be answered locally, but they reflect the centrality of community concerns when tech stewards consider technology and its potential.

Connectivity and proximity. Increased connectivity produces a kind of virtual proximity. "Always on" translates into "always there." Once the natural limitations of the physical world are removed, the potential for being together appears boundless. This may be alluring but it can also disrupt the rhythm of a community that depends on its members taking a break from community engagement. How much contact is enough and when is it too much? When is connectivity a source of cohesion and when a source of overload or even fragmentation? If some members use technology to reach high levels of connectivity, can tech stewards avoid inadvertently creating second-class citizenship for those who are not so well-connected? For tech stewards, connectivity is not a matter of technological capability, but one of rhythms and cohesion.

Shifting boundaries and peripherality. Increased connectivity also opens new peripheries by allowing many people to observe without committing, for example by reading a blog, a discussion, or a wiki, but never commenting. What are the effects of such peripherality on a community? Does it add dynamism or disrupt the intimacy of a group learning intensely? As technology affords the power to control detailed access to specific content and interactions by both members and non-members, the management of boundaries becomes a necessary part of the learning agenda for tech stewards.

New modes of engagement. The evolution of publishing and interaction tools enables both stronger individual voices, through self-expression tools like blogs, as well as stronger group manifestations, through collaboration tools like wikis. But again the successful use of these capabilities for communities depends on the practices members develop around them. When does a group blog make sense and when individual blogs? When can an individual speak for the community? What are the practices that need to be learned about modifying each other's writing on a wiki? When does a community need to acknowledge the contribution of individuals to a communal product? How do these new media change the role of moderation? When interactions live on as products and products live through interactions, communities need new practices of engagement.

Creative reappropriation and community voice. We have noted that the hypertext nature of the web makes it possible for communities to create entire resource collections through links and tags. How does this affect a community's sense of ownership of its "repository"? How can the publication of its own products be integrated into a system of pervasive cross-references? Being a productive node in this hypertext world requires new practices of participation and in some cases, a new sense of community voice.

Transparency in a socially active medium. The digital medium can produce enormous amounts of data about participation, contribution, the use of resources, and about members themselves. Techniques for mapping networks enable new ways of visualizing the structure of communities. Digital footprints can provide much information about individual behavior. Polling can reveal the "mind" of the community. The potential transparency afforded by these "meta-representations" is not an unqualified bonanza unless communities learn to use them appropriately. When do issues of privacy and individual rights conflict with the potential for community insight? When does a reputation system pit members against each other to the point of damaging a community's sense of collegiality and therefore its learning potential? How can a community use polling to create collective views while avoiding majority rule and groupthink? Here, as in many decisions, it is preserving and fostering the ability to learn together that should guide a tech steward through these balancing acts.

Dealing with multiplicity. What we've called homesteading of the web, combined with expanding modes of self-expression and collaboration, produces an abundance of forms of participation. This richness combined with limited attention exacerbates the tensions inherent in the individual/group polarity. Our communities will have to deal with the competition generated by the multiplicity of places where individuals can congregate to discuss any given topic. The personal self-expression channels of individual members can also seem to compete with the gathering places that their communities sponsor. Will members still want to contribute to the community in the same way once they focus on their own blogs? As multimembership enables people to connect with each other in more and more places on the web, will they still feel an allegiance to one community in particular? Will any of this matter as long as the diversity of locations and channels can be linked and aggregated dynamically for different purposes? It will be crucial for communities to learn to engage with this multiplicity and partiality of participation, and to leverage its dynamism rather than treat it as a threat.

2. Making new communities possible

The second area in our learning agenda concerns the potential that digital habitats present for enabling new groups of various kinds. Many of those groups will become new communities, which would not have formed without technology. We might not even recognize them as communities until we look carefully at what technology enables. These technology-enabled communities present particular challenges that require the development of new practices.

People finding each other on a wider scale: size and meaningful engagement

A socially active medium provides new avenues for people to find each other across all kinds of boundaries. People can come together who would not meet otherwise—or even know of each other's existence. The ability to find each other and interact more easily across distance means that people can form communities that are potentially very large. We have already seen many examples of very large, very successful communities of practice that have changed our expectations in this regard. This includes some of the communities mentioned in this book, such as the MPD community, CompanyCommand, the Solucient customer communities, and the Webheads. Understanding in what ways these groups have been able to become meaningful communities (and how they are supported over time) is part of our proposed learning agenda.

Size is certainly a challenge for communities. As technical barriers to group size disappear, we will have to be more systematic in exploring issues of scale and limits to growth as well as the practices to address them. Finding the limits to meaningful engagement in a chat room, a phone bridge, an email list, or an RSS feed is not a technical problem; it is a matter of meaningful social interactions. Scaling up community size is not a linear progression from small to large. It can entail subgroups, peripheral participation, and geographic divisions. Many large communities, such as CompanyCommand and Solucient, spawn and nurture topic-oriented subcommunities where participants experience high degrees of focus, intimacy, and practical learning. Rather than subgroups, the MPD community has a very large periphery of members who benefit deeply from their participation even if they rarely or ever say anything at all. These are the types of practices we will have to develop to cope with the large-scale communities that technology enables.

Digital habitats as community catalysts

To the extent that technology creates new spaces for togetherness, it can itself become a catalyst for community formation, whether or not this was the intention. People posting on a discussion forum, contributors to a wiki, or commentators on a blog are not necessarily a community, but the potential is there. When such a potential is present, understanding the paths from technology to community is a key element of stewardship, especially because there may not be any community leaders yet to recognize and convene the community. In a twist on the theme, this is technology stewardship in reverse: rather than finding technology for community, it is finding community in technology.

Even if a community does develop, it runs the risk of continuing to be defined largely by its original technological space: a listserv, a blog, a conferencing system. When membership is defined by participation in a technology space, it may take new skills to give it an identity that is not defined by technology. For instance, articulating a domain and discovering activities for learning about the practice that goes with it allows a group to find an identity that is defined by its learning rather than a specific tool. Such a group can then start to move across technologies to find the best configuration to serve its learning.

Access to living practice through virtual presence

As communities become global, supporting meaningful engagement at a distance is becoming an imperative. Interaction tools offer the potential to learn together as practitioners even when distance prevents participants from actually practicing together. Participants in a web-based forum often do not engage in practice together in the forum, but their shared background in practice makes their exchanges very meaningful. This convenience cuts both ways. For instance, when anyone can post on an online forum or on a blog, it makes for informal, open-ended conversations that can be very useful to practitioners. However, the very openness and informality of some of these conversations make it easy for a vocal minority to take over and leave a large silent majority wondering whether participation is worth their time. How do we judge whether the active contributors are energizing the community or boring it with another round of grandstanding? How do we recognize their commitment to the domain or their actual experience of practice when the channels of communication are easy to access but detached from practice? When technology-mediated interaction makes it easier to hide one's level of personal experience and commitment to learning, this can lead to group dysfunction.[1] What level of reflectivity, cultivation, or backchannel conversations should a tech steward or any other community leader encourage? We need to become much more attuned to the possible pitfalls of digital habitats and to the cultivation practices that can alleviate potential problems.

The more successful large-scale online communities mentioned in this book seem to be able to keep a strong focus on practice and foster productive learning. In cases such as MPD support and CompanyCommand, this focus seems due both to the model of the leaders and to the urgency of the domain. CompanyCommand also makes use of tools such as video to bring practice to life. New interaction tools are an exciting medium for peer-to-peer learning and many technological developments have promising applications for communities, such as cheap teleconference bridges, virtual meetings spaces, desktop videoconferencing, and even avatar-based immersive environments like Second Life. Putting a community's shared practice at the center of a shared experience requires more experimentation and expertise than just having online access to each other. The technology is ahead of the practice of learning together and there remain many challenges. A simple problem like dealing with time zones is still an obstacle; and we do not have good solutions to the pervasive issue of multiple languages. Facilitating learning through virtual presence is central to the learning agenda of technology stewardship if we are to enable new kinds of communities.

Complex geographies of identity and domain-based relationships

The ease of finding others based on common interest not only has an effect on size; it also yields more specialized groups. The focus on very specific topics in digital habitats creates very narrow contexts of participation, each of which reveals a thin slice of the person— that part of their identity that is related to the specific domain. This is true of all contexts of participation to some extent, but online participation, especially in writing, increases the phenomenon because community-specific interactions and contributions are often all that participants see of each other.

Sometimes these "thin" connections are appreciated because they enable a focus on content; sometimes they feel like a poor substitute for deeper relationships. Technology can expand as well as limit the images that people are able to form of each other. A digital footprint can reveal information about a person's multifaceted identity beyond what would naturally be revealed in the course of daily interactions. Communities can learn a lot about their members—by "Googling" them, for instance—and digital footprints can provide even more insight into identity than meeting the same people face to face.

We still do not have a very good understanding of how the multifaceted nature of human identity plays out online. When is an orientation to relationships called for as a vehicle for

---

1. Clay Shirky, "A group is its own worst enemy." http://www.shirky.com/writings/group_enemy.html, Clay Shirkey's Writings About the Internet June 30, 2003 (acessed March 31, 2007).

trust and learning and when is it a distraction? When are the depth of domain-focus and the depth of relationships in synergy—and when in conflict? The advantage of domain-focused channels of communication is that a member is not distracted by other aspects of a person's life and identity. Who cares what else a member is in other contexts, outside of the community, if they contribute to meaningful exchanges in the community? That a member trusts and respects another member as a contributor in one community does not mean they should trust that member in another community, or entrust them with their children, or ask their advice about personal finances.

At the same time, knowing only a thin slice of others can make it more difficult to interpret where their thoughts and statements are coming from. What does it take to know each other meaningfully within a given context and across contexts? Does the multiplicity of contexts change the technology implications of an orientation to relationships? It used to be that identity was established by a personal web page on a community site or a picture gallery. But now members have their own broad digital footprint spread across the web. To what extent is the need to understand each other across contexts a matter of personal preference? The attention we can devote to experiencing these widely distributed digital footprints may be limited by time constraints or personal choice. Do you care that I read all about you across the web, but you haven't even peeked at my bio? Reaching enough mutual commitment to learn together is always a delicate process that includes knowing enough about each other and building trust, technology introduces new twists to this challenge.

Seeing the social in the technological

While there is no question that digital habitats can give rise to new communities—by connecting people across time and space, by creating new spaces for engagement, by revealing affinities for shared domains, and by providing information about people—we need to make a clear distinction between the technology and the social conditions and processes that bring a community together. The two are closely intertwined as technology increasingly becomes a social medium. But because technology is often more visible than the social conditions it manifests, it is easy to confuse the two. Just because the technological container remains, it does not mean the community is still functioning and alive. Conversely, the disappearance of a website does not mean that the community that used to live there has died. No matter how technology-enabled they are, communities remain social entities and it is by enabling social processes that technology contributes to the emergence of communities. Tech stewards must learn to recognize the social processes that technology enables and understand how to support these processes as a way to foster the emergence of meaningful communities.

3. Stretching our very notion of community

The third area of our learning agenda is the recognition that the interaction of technology and community affects both to the point of challenging our assumptions about them. It puts a social spin on technology and it opens new questions about who we want to be, who we can "be with," and how we can connect around what we care about. Insofar as it reconfigures fundamental polarities, it stretches our understanding of a community of practice and its difference from a network in terms of size, stability, diversity, boundaries, and modes of engagement, as well as personal and collective identities.

Proto-communities: emerging patterns of communities and networks

We increasingly see the emergence of what could be called proto-communities—amorphous, networked collections of people, places, artifacts, and activities that are slowly developing an identity as a community.

A community of practice represents an intention—however tacit and distributed—to steward a domain of knowledge and to sustain learning about it. A strong community is a group where that intention is well understood as a joint enterprise, no matter at what scale or with what level of interaction. This implies a high level of identification with the domain as something that connects the members and their orientation to practice. A network by contrast entails connectivity, with nodes and links, but not necessarily an intention to hold a domain or identification with a joint enterprise.

As technology allows networks and communities to overlay and stretch each other in new ways, we need to both keep the distinction clear, to be able to maximize the benefit of each form, and at the same time see that they morph into each other.

Several trends contribute to increased interweaving and blurring of community

The NPtech community: What is in a Tag?

Beth Kanter had no intention of starting a community. She wanted something useful to say on her blog. She noted that the NPtech tag had attracted a lot of attention and that it was pointing to a lot of resources for people interested in technology for non-profits—so many resources, in fact, that it was getting hard for anyone to follow. So, she decided to produce a weekly summary. And she tried to get in touch with new users of the tag, drawing them further into the conversation. The summary was soon noticed and started to draw wider attention. Today Beth's blog is one of the key nodes in an active network of people interested in technology for non-profits. Many communicate with her directly, on her blog, or by email; others communicate through their own blogs or on various forums; many still participate only by tagging relevant resources. More recently, Twitter has become a mode of interaction through which people point out relevant resources to each other. The NPtech tag has clearly revealed a domain and a proto-community is now emerging, slowly gathering an identity, building relationships, finding ways to address issues of practice, and migrating across technologies.

Giving a single URL for such a highly distributed proto-community is impossible or misleading, almost by definition. There is a description of the community at http://www.nten.org/blog/2007/03/20/nptech-tagging-community. Beth's blog is at http://beth.typepad.com/beths_blog/2007/11/nptech-tag-su-2.html. Some information can also be found at http://nptech.info and under the tag itself at http://del.icio.us/tag/nptech or on other platforms such as flickr.com or on youtube.com.

and network processes: the homesteading of the web, which multiplies relevant locations for a domain; generalized self-expression, through which distinct voices emerge; dynamic boundaries, which are defined by activity and popularity; and a socially active medium, which connects all this through real and potential links. Sustained attention to important domains can therefore be highly distributed. It is carried out by groups and individuals who care about the domain through conversations in various places, on blogs, on wikis, and through tags.

Where communities end and networks begin is not always clear in such a dynamic geography with rapidly shifting boundaries.

For instance, conversations around blogs can generate emergent communities whose boundaries are dynamically defined by participants as the conversation evolves.[2] These communities take different shapes.

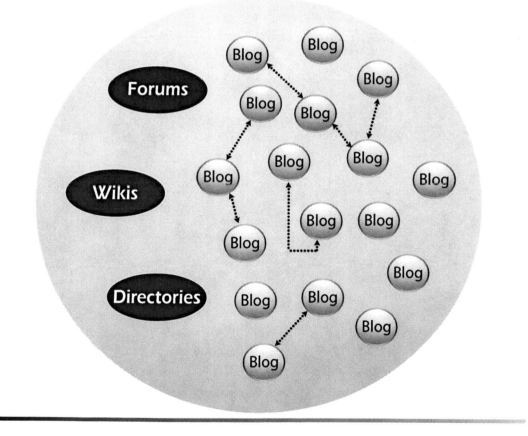

Figure 12.1. Bounded Communities

2. For a more detailed description of these graphics and the types of communities that can arise in the blogosphere, see Nancy White, Blogs and communities: A new paradigm for online community (2006). First published in the Knowledge Tree, http://www.fullcirc.com/weblog/2006/12/blogs-and-community-launching-new.htm (accessed April 27, 2008).

Some blog communities sit inside a specific platform, as illustrated in Figure 12.1. Blogs may be combined with other tools like discussion forums and member directories. These configurations allow lots of dynamic interplay of communal and personal spaces, even though they are controlled by the site host and have a strong sense of external boundary. An example of this is the March of Dimes "Share Your Story" community, where parents who have babies in neonatal intensive care units offer support and advice to each other. They are invited to introduce themselves in the community space, join discussions on various topics, and start a blog to chronicle their experience.[3]

Some blog-based communities are focused on one prime blogger and a constellation of readers and commenters, some of whom have their own blogs. The hub created by the primary author gives the author much influence over the interactions and topics; there is a high degree of identification with the author and a lot of peripheral participation (Figure 12.2). Beth Kanter's blog certainly plays a role of this type as part of the NPtech tagging community described in the box above. Since knowledge management expert David Snowden started his blog in the summer of 2006, his provocative style and sharp

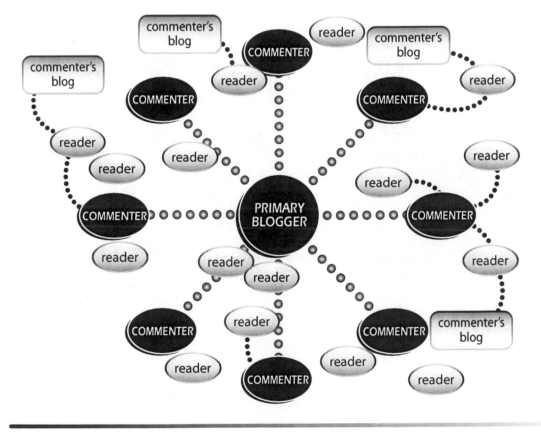

Figure 12.2. Community arising around a primary blogger

3. March of Dimes' Share Your Story http://www.shareyourstory.org

Chapter 12. A learning agenda

insights have generated a lot of comments on other blogs.[4] As the back-and-forth interactions reverberate through the blogosphere and beyond, a collective inquiry is being pushed forward in the gray zone between community and network—with rapid diffusion in a network format, but enough identification to suggest a proto-community.

Blogs that connect and form communities based on a shared topic or domain have a network structure, showcase the identity of multiple bloggers, and have decentralized control. But again over time the back-and-forth among related blogs and the identification with the domain create a community-like structure. Such an evolution is visible in the Edublogger community (individuals interested in the use of blogs and other technologies in education as described in the box on the Edublog RSS in Chapter 5). This is a large and growing group of bloggers with significant regional identification, probably because of the local nature of educational issues. The Edubloggers have asserted their identity as a global community, and now we are seeing sites that endeavor to specifically aggregate content across individual Edublogs.

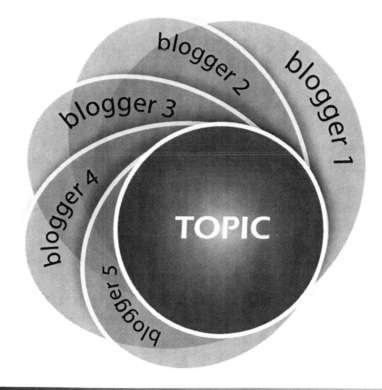

Figures 12.3 Topic-centric communities

---

4. Dave Snowden's Blog, http://www.cognitive-edge.com/blogs/dave

In this emergent type of community the notion of boundary is less a matter of a clear "in vs. out" than a matter of evolving shape, which is defined by dynamic relationships between the voice of individuals and the emergence of the community. This rapidly evolving mix of voices, interactions, resources, and links allows communities to form and domains to be identified without well-defined or stable boundaries.

Sometimes a domain requires the focus of a well-defined community of practice, with a clear identity, voice, and boundaries; sometimes it is best served by a less unified proto-community. For instance, there may be many groups interested in astronomy as a hobby, but the scientific community needs a clearer identity because of the nature of its enterprise, which requires that members be accountable to each other's work in a systematic fashion. There may be many informal support groups for parents who have children in neonatal intensive care, but in the case of the "Share Your Story" community mentioned earlier, the March of Dimes has decided to support the emergence of a number of well-defined, bounded sub-communities in which members care about each other personally. For instance, in the group for women who are pregnant after having had a previous high-risk pregnancy, members share tips and experiences, as well as information about doctors and procedures. But the identity of the community is also based on personal relationships: members follow each other's stories; they check in every day; and if a member does not check in, someone will call her to see what is happening.

In summary, the need for a proto-community to strengthen its identity can be prompted by the various aspects of learning in community introduced in Chapter 1: the nature of accountability to the domain, the importance of sustained relationships, or the desire to align practice. But even with a strong identity, communities can remain dynamically spread over multiple "locations" and ways to connect. For a good example of one of these highly successful "multilocation" communities, see the box on the story of the Webheads in Chapter 4 and the box on the Staffordshire Best Practice community in Chapter 11.

Emerging practices of stewardship: balancing network and community processes

Proto-communities have always existed as people in various places have developed similar interests and practices. What is new is that the constitutive elements of these proto-communities are now directly findable and linkable through a common infrastructure. People who do things of mutual relevance can find each other and connect—whether transiently or to form a community. As a result, the process of caring for a domain is increasingly a mixture of both community and network processes. Communities and networks are not distinct structures. Groups can have elements of both. Rather they are distinct structuring processes that work in complementary ways in the same geographies. Caring for a domain requires a balance between community-oriented processes such as promoting collective identities around the shared domain, on the one hand, and network-oriented processes such as connecting, aggregating, remixing, and brokering contributions across locations to strengthen and expand emergent networks, on the other. It is the interplay of these two types of processes that can create the mix of focus and dynamism necessary for learning together.

For tech stewards, the learning agenda is to recognize these community-network hybrids and help them take shape and realize their potential without imposing on them a limited, obsolete, or romanticized view of community or network. What are the networking and community affordances of different tools and technologies? When a network lacks self-awareness, can technology be used to build a collective identity? When a collective identity has become too closed and inward-looking, can technology be used to open up communities to new network connections?

Learning between the old and the new

There is a fine balance between updating our understanding of community in light of technology developments and doing justice to what we have learned about community, network, and the social nature of being human. Calling every online space or grouping a "community" is not useful. We can't help but cringe a bit when we hear people call Wikipedia a community of practice, for instance. While there will be communities of practice around clusters of pages and around editorial responsibilities, Wikipedia is a space with many constituencies, forming a group of people too amorphous to talk about shared identity, practice, and community.[5] At the same time, it is important to see notions like communities of practice and networks as evolving perspectives on social learning: it is their conceptual dynamism, not a frozen idea, that makes them useful. Part of our learning agenda is to follow these developments with an open mind while still keeping enough of a sense of what matters about human learning to be able to generate useful practices.

Technology stewardship: a latent community

We have framed this final summary of our findings about the interactions between communities of practice and technologies as a learning agenda. We used the term to emphasize our belief that learning to marry technology and community is a challenge that brings many players together. A learning agenda is a good focus for a community to form around. Actually, we already feel part of an emerging community interested in technology stewardship, but it is a good example of one of those diffuse proto-communities:

---

5. James Paul Gee suggests that we talk about a "semiotic social space" rather than a community in such cases. A semiotic social space is defined by the informational and participatory structure of the space itself rather than by the group of people associated with it. See J. P. Gee, "Semiotic social spaces and affinity spaces", in Barton, D. and Tusting, K. (eds.), Beyond communities of practice: Language, power, and social context, (Cambridge University Press, 2005). Of course, given the distinction, whether a semiotic social space does or does not reflect or enable a community, whether it should, to what extent, or in what ways—all these remain interesting questions to ponder for both theoretical and practical purposes.

highly dispersed although highly connected in pockets, active, passionate, and in the process of discovering itself. When we mention the notion of technology stewardship and its challenges, people often nod in recognition. There is an emergent practice here. It is an identity that is beginning to form.

To what extent our recognition of a shared learning agenda will translate into a well-defined community, we cannot know. We hope that our writing will spark both fleeting and sustained conversations. If this book can begin to frame these conversations, we will be satisfied. The prospect of strengthening an experience of community among people interested in digital habitats for communities is exciting for several reasons:

- There is value in comparing the challenges and practices of technology stewardship across organizational, social, linguistic, cultural, and technical contexts.

- There is value in bringing a community perspective to technology development. We have found it productive to balance cleverness about technology with a deep commitment to the life of the communities that can benefit from the support of clever technology. This community sensitivity entails a series of conversations among tool builders, people interested in the sociology of the Internet as a place of learning, tech stewards embedded in communities, and community members.

- There is value in finding a voice to address some of the broader issues that affect communities and their learning potential, such as the digital divide, questions of intellectual property, or the openness and neutrality of the Internet.

- Finally, there is value in clarifying and cultivating the role of technology stewardship so that those who engage in this practice develop a productive sense of their own identity. Technology stewardship is not merely about technology, technical support, or even user support. Its fundamental aspiration, as we see it, is to realize the evolving learning potential generated by the interplay between technology and community.

All this is happening at a time when the need for learning and community is intensifying. A number of contextual trends give some urgency to our work:

- Organizations in the private and public sector are discovering the power of communities and peer-to-peer processes; these organizations struggle to understand the implications of all this for their work and their place in a new environment of collaborative technology and networked interactions.

- Humankind is facing profound environmental, economic, cultural, and political challenges that require new kinds of communities to learn together. A principle of social learning theory is that our communities have to match the problems we are addressing in size and complexity.

- Globalization forces us to work across boundaries of all sorts: geographic, disciplinary, organizational, and cultural boundaries are the challenges we next encounter once we can easily reach out across time and place. This challenge is clearly one of learning to participate in reconfigured geographies of identity.

In retrospect, what started out as an examination of tech stewards "doing their thing" for their own communities has become a new window on something that looks like a social movement, producing innovation on a much larger scale. The daily activities, concerns, and aspirations of tech stewards are viewed in the context of a broader momentum of inventiveness whose consequences for the world are yet to be determined. Although technology by itself cannot assure positive outcomes, ideally the ultimate effect of careful stewarding is an increase in community and in learning capacity. And in the aggregate, careful technology stewarding across many communities is bound to have such broader effects at a time when they are urgently needed. In the last analysis, this is why we find this topic so appealing. Questions of broader effects are especially promising if we consider that this perspective on technology stewardship is not limited only to specialists who take on an explicit role. It is part of basic citizenship in regard to learning in a technology-rich world. It is something we can all care about—whether and how our participation in our communities and their digital habitats contributes to the learning of our small planet.

Glossary

To make this book more widely accessible to readers with different kinds of experience, this glossary defines key technical terms. These definitions are aimed to increase understanding, rather than being the last word. **Bold** words within glossary definitions are see-also terms. Other resources for terminology are at the end. Most of the terms included here are technology-related.

Access lists/Access controls: Controls that limit access to a resource, such as a discussion or a tool. Sometimes called "permissions." Typically a responsible person will grant access or permit an individual to perform certain actions, such as being able to read, post, modify, or delete, etc.

Aggregation: The process of automatically gathering material from multiple websites (blogs, news sites, podcasts, etc.) and displaying them on a single web page. This is typically done by subscribing to a **feed** and then reading them with an **aggregator**, such as an **RSS** reader.

Aggregators: Software tools that bring together material from different sources and display them on one page. For example, a blog reader allows a user to subscribe to posts from different blogs and then read them on one page. There are also aggregators that pull together diverse content, including RSS feeds from blogs, photo services, "microblogging" services like Twitter, and instant messengers. The latter are sometimes called "friend aggregators."

AJAX: Shorthand for **A**synchronous **J**avaScript **a**nd **X**ML. A programming technique for interactive web applications. Using AJAX, some of the request processing between server and client (such as a browser) is handled on the client side. Thus the server does not need to reload the whole web page each time the user requests a change. This increases the speed and interactivity of the page.

Always on: Refers to a service or method of access to the Internet that stays on, as opposed to the intermittent connections through dial-up modems. Cable and broadband connections are always on.

API or Application Program Interface: A standardized, published method for communication between systems. An example of integration through standards (where standards may be local agreements as to how systems will communicate). Some APIs may be open and public and others proprietary and available only to licensed users.

Application sharing: Systems that allow you to broadcast what you can see on your computer to a larger group. Useful in demonstrating how to use a computer program. Also called **desktop sharing** or screen sharing.

Archive: Files or material from online interactions that have been closed for participation but kept as a record of the interaction. **Chat** archives are often called chat transcripts.

Asynchronous interaction: Different-time communication (as opposed to synchronous, "real-time" communication like a **chat**). Online discussions occurring independent of time or location. Participants send or post messages to a central location, such as a **discussion board**, where they are stored for retrieval by other participants. Examples of asynchronous interaction include web-based discussion forums, wikis, blogs, bulletin boards, and email.

ASP or Application Service Provider: A vendor that provides a software service including the software, hosting, and usually support. For example, Blogger offers blog software and hosting. This is useful when you don't have your own server or want to be free from doing basic support tasks. The downside is you often have less ability to configure and tweak the software. Costs range from free to very expensive. ASPs that provide the service for free often support it by including advertising.

Attachments: A file from another application appended to an email, discussion post, or other communication. Examples of attachments: spreadsheets, documents, photographs, programs, and so on.

Audio recordings: Digital audio files, such as voice or music. These are created and stored in specific file formats, such as .MP3, .WAV, and others. They require the appropriate software (or media player) for listening. Some media players handle a variety of file formats.

Avatar: An icon or representation of a person in an online environment. It can be a picture in a discussion forum or a moving character in an online game or 3-D **immersible environment** like Second Life. Some avatars are quite fanciful and may serve as an expression of some aspect of a user's personality or alter ego.

Backchannel: Personal, private communication (email, instant messages, private **chats**, phone calls) between one or more individuals, as opposed to interacting in a public discussion forum. Backchannel communication is usually informal, can have a big impact on online interactions, and can sometimes become public or visible to a larger group. Backchannels are sometimes used to "whisper" to others. Moderators and facilitators use them to coach or talk privately with participants.

Behavioral parameters: Used for rating systems where members can rate contributions of others. These can enable online community leaders or moderators to define "desired" behaviors that then become visible to the community. Participants with high ratings can be rewarded with special privileges and those with low ratings restricted or even banned. A rating system is a simple form of a reputation system. See also **rating tools**.

Blog: A blog (short for "weblog") is a website with commentary, news, opinions, and the like, often from one person. Some blogs have multiple contributors. Some also allow **blog comments** from readers. New materials (posts) are displayed chronologically, often with new posts at the top. Blogs are created with programs that allows easy web publication without the hassles of HTML editing and having to use a separate program to upload your work. Look and feel can be set with templates or customized. Blog software is becoming so flexible that you can create entire websites that don't look like "traditional blogs."

Blog categories: A set of defined keywords that are used on a blog to classify posts so they can be viewed together, in one stream. Categories are a controlled vocabulary for tagging associated with that blog.

Blog comments: Some blogs allow readers to add comment to individual blog posts. Comments are used for things like discussion, expressing other opinions, providing more information, and correcting errors.

Blog trackbacks: A method for blog authors to request notification of links to their posts by other bloggers. This is one way that bloggers can have discussions with each other. A person posts a blog entry with commentary about someone else's post and includes a link to the original. The original author is then notified with a trackback. Some blog systems automatically post **trackbacks** with a post so readers can find all the sites where it has been linked. Also called pingbacks.

Blog readers: Software to gather different **RSS** feeds together for presentation to the user in a suitable form for browsing. Features include locating the **feeds**, marking or storing individual items, and importing or exporting a file that describes the feeds you want to read. See also **aggregators**.

Blogroll: "A list of recommended sites that appears in the sidebar of a blog. These sites are typically sites that are either on similar topics, sites that the blogger reads regularly, or sites that belong to the blogger's friends or colleagues. The term 'blogroll' also evokes the concept of political logrolling (when legislators promise to vote for one another's pet bills) — which is not unlike bloggers' habit of reciprocating links by posting links to blogs that link back to their own blogs." <u>Source</u>: Social Signal, http://www.socialsignal.com/blog/alexandra-samuel/web-2-0-glossary

Blogosphere: The collection of interrelated blogs, representing a global conversation.

Bookmark: (*verb*) To mark a document or a specific place in a document for later retrieval. Most web browsers support a bookmarking feature that lets you save the web page address (URL) so you can easily revisit the page at a later time. (*noun*) A marker or address that identifies a document or a specific place in a document. See also **social** or **shared bookmarking**.

Bulletin board: One of the names for web-based online conferencing spaces. Bulletin boards are asynchronous tools and can be organized in linear or threaded formats. Other names for online conferencing include forums, discussions, and **newsgroups**.

Buddy lists: A user-specified list of individuals who are "known" to a user of an **instant messaging** system or a **VoIP** system like Skype. A buddy list makes it easier to reach someone or gives him permission to contact you or see when you are available. Also known as contact lists.

Calendars: Calendaring tools allow communities to keep individual and group schedules of events. They can be standalone applications or they may be integrated with individual users' tools such as Outlook. See also **shared calendars**.

Categories: See **blog categories**.

Chat: Same-time (synchronous) text interaction over the Internet. Typically fast moving, chat can be used for large "auditorium" events where there are presenters and an audience, for smaller group work meetings or social interactions, or for small, one-on-one sessions. Some chat applications now integrate voice as well as text chat. There are many variants but no dominant standard for chat. Examples range from several people who use a common **instant messaging** (or **IM**) system, to online chat rooms with hundreds of people listening in, to **IRC** channels.

Computer-mediated communication (CMC): Communication via online tools such as email, web pages, online interaction, or conferencing.

Content management systems (CMS): A repository platform that provides methods and tools to capture, manage, store, preserve, and deliver content across a single or distributed environment. Many such repositories have features to search, tag, or rate the files that are stored in them. A CMS may include distributed or centralized organizing and **access control** mechanisms, and is often integrated into a portal platform.

Cross-platform issues: Programs run on different operating systems. A Windows-only program will not run on an Apple or UNIX system. Standards are also implemented somewhat differently for various browsers. A web page that looks good with one browser may not work properly in another. Cross-platform issues are important in selecting and using systems and tools for communities that are accessed with many different kinds of computers and browsers.

Database: A structured collection of data that can be searched, retrieved, and updated by one or more users.

Desktop sharing: Using web-based software, desktop sharing allows one user to show her computer's desktop to other users. This can include letting the other users see and even use the software installed on the first user's computer. See also **application sharing**.

Directory (member directory): A list of people who are registered users on a website or platform, who use a particular software system, or who are members of a community. Directories can contain everything from computer usernames to **individual profiles** with contact information, photographs, and links to contributions.

Discussion board: Web-based discussion boards, also known as forums or **bulletin boards**, are a classic tool that allows groups to conduct asynchronous, written conversations online. Participants can visit the conversation space and see a whole discussion as a series of posts. They can contribute to the conversation by posting a new message. Each message includes a date/time stamp and author information. Because they are a traditional tool, discussion boards tend to have many other tools attached to them (from **directories** to **wikis**, etc.)

Distributed: Refers to a group of people who are not in the same geographic location. The word distributed is often used in conjunction with teams, as in a "distributed team." It also refers to software, data, and other computing objects that are spread out between processors.

Document management: "The computerized management of electronic as well as paper-based documents. Document management systems generally include an optical scanner and a system to convert paper documents into an electronic form, a database system to organize stored documents and a search mechanism to quickly find specific documents. Key to document management is an understanding of the potential uses of the source material and a protocol for identifying content elements (such as keywords and other characteristics of a document.)" Source: Webopedia, http://webopedia.com/TERM/d/document_management.html

Document sharing: A form of **file sharing**. People who are collaborating often share documents, including text, spreadsheets, presentations, etc. Some document sharing systems allow real-time editing by several people. The practice of sharing documents by email is an informal way of collaborating. **Document version control** is important to track revisions.

Document version control: Information about the version of a draft that is relevant to the development of a document, as distinct from the information that is appropriate in a library. Tracking versions of a document as it evolves or goes through the editing process.

Download: (*verb*) To copy and transfer files from a website, web space, email system, or another computer to the user's computer hard drive for offline use. (*noun*) A file that is so transferred. The opposite of **upload**.

Email: Short for "electronic mail". Messages transmitted over electronic communications networks. Some electronic-mail systems are confined to a single computer system, intranet, or network, but others have gateways to the Internet, enabling users to send electronic mail anywhere in the world. (For some basic tips on effective email, see http://www.darwinmag.com/read/100101/ecosystem.html.)

Email list: A group email function that sends or "broadcasts" a single email message to a group of people. Email lists use a variety of software tools with features that may or may not include multiple modes of participation (for example, individual messages or daily digests containing all the day's messages), automatic archiving of posts on a website (such as Google Groups and Yahoo! Groups) and varying levels of moderator control.

Expertise systems/Expertise locators: Software tools to allow people to find other people with particular expertise. These can be deployed within an organization, or externally. Some **social networking tools** are also expertise locators.

F2F: Shorthand for "face-to-face," to convey offline interaction where people meet in person.

Feeds: Also known as web feeds or **blog** feeds. "A web feed is a document (often XML-based) which contains content items, often summaries of stories or blog posts with web links to longer versions. News websites and blogs are common sources for web feeds, but feeds are also used to deliver structured information ranging from weather data to 'top ten' lists of hit tunes... More often, feeds are subscribed to directly by users with **aggregators** or **feed readers**, which combine the contents of multiple web feeds for display on a single screen or series of screens. Some modern web browsers incorporate aggregator features. Depending on the aggregator, users typically subscribe to a feed by manually entering the URL of a feed or clicking a link in a web browser." Source: Wikipedia, http://en.wikipedia.org/wiki/Blog_feeds

Feed reader: A tool that collects all the **feeds** a person has subscribed to and puts them into an organized, readable form on the desktop or in an Internet browser. Also called a newsreader or **aggregator**.

File sharing: "The term file sharing refers to the sharing of computer data or space on a network. File sharing allows multiple users to use the same file by being able to read, modify, copy and/or print it. File sharing users may have the same or different levels of access privilege." Source: Findlaw, http://technology.findlaw.com/law-technology-dictionary/file-sharing.html

Firewalls: Systems that protect one or more computers or networks from harmful access. Sometimes an organization's firewall will restrict access to a specific website or prevent the use of a specific kind of software.

Folksonomy: A portmanteau word combining "folk" and "taxonomy" that refers to the collaborative but unsophisticated way in which information can be categorized on websites. Instead of using a centralized form of classification or controlled vocabulary, users are encouraged to assign freely chosen keywords (called **tags**) to pieces of information or data, a process known as tagging. A folksonomy is a user-generated taxonomy. Examples of web services that use tagging include Flickr and del.icio.us.

Frequently Asked Question or FAQ tools: Questions that users commonly ask about a system, community, tool, or the like, with answers that apply within that context. Tools to organize and deliver FAQs.

Friend aggregator: There are several kinds of friend aggregators. Some people say that social networking sites like MySpace and Facebook are friend aggregators. Another kind is a **feed** of web pages, photos, videos and music that your friends are sharing, such as FriendFeed and Socialthing.

Geomapping: The process of creating maps for the web, combining places (locations) and other information from a variety of sources. Some geomaps also contain time information, such as when a storm front is predicted to be where. Many **mashups** are combinations of maps and data. You can get map **widgets** for your website, or create your own geomaps with tools like Google Maps and the Yahoo! Maps API. Also called web mapping.

Geotagging: The process of adding geographic tags, usually longitude and latitude, to websites, web pages, images, and other objects for which location is important. Used with **geomapping**.

Host (or web host): There are a variety of meanings for this word. From a technical perspective, it is a computer acting as an information or communications server. The word host is also used to define the role of someone who is providing and/or facilitating an online space.

Hosted application: A software application that is provided on someone else's server. Generally, hosted applications include support as well. See also **ASP**.

Hyperlink (or link): A mechanism in a document that links text in one location to another. Clicking on the hyperlink opens the linked document or location.

Individual profile: A page with contact and descriptive information about an individual that is visible to the whole community. Profiles vary in terms of information shown and how access is controlled. System administrators may have access to some information that others do not. Profile pages are often collected in a **Directory**.

Immersive environments: Computer programs that create a 3-dimensional illusion which suggests that a user is "in" an imaginary or virtual world. Examples include Second Life, some online games, and virtual reality spaces.

Infrastructure: Refers to the underlying hardware, software, and networking systems and tools needed to support an online activity or community.

Instant messaging (IM): Instant messaging allows users to send small amounts of text in real time from one computer to another, often accumulating messages in a shared text window. Many instant message programs allow **chats** with a small group of known participants in addition to private one-to-one conversations. Instant messaging can be part of a larger platform, so that it is integrated with presence and other tools such as **individual profiles** and other resources specific to one community or one platform. Conversely, instant message programs have become their own platforms, providing **buddy lists**, **VoIP**, **presence indicators**, etc.

Intranet: A network of websites and servers that is available only within an organization, usually behind a firewall. Often organized to provide a consistent look and feel.

IRC (Internet Relay Chat): A **chat** system available to anyone with Internet access. Live discussions take place in IRC channels, and there are thousands of active channels all the time. You can also open a new channel for a chat and then invite specific people. To join an IRC conversation, you need an IRC client or a browser (or plug-in) that supports it.

Look and feel: A website's highly usable, appealing look makes it a pleasant place to come to. Some systems focus on making their appearance highly customizable and others on having something appealing right out of the box. Templates are a method for providing commonly used customizations.

Lurker: Someone who reads in an online interaction space, but rarely or never posts. Lurkers are also known as "readers." When they do post, they are said to be "delurking." Depending on the purpose of the interaction space, the facilitators may try to engage "readers" to begin responding and posting. "Lurker" sometimes carries negative connotations, so in some settings, using the term "reader" is advisable. If the purpose of the space is to share information with a wide audience, lurkers are not necessarily a problem. Readers provide an audience and they provide page views—they are an influence, albeit unseen and sometimes hard to understand. But if the purpose is to generate and share new information, it's useful to encourage everyone to post. Readers can be converted to posters.

Malware: A portmanteau word combining "malicious" and "software." Malware is software installed on the user's computer without explicit permission. Examples include computer viruses, worms, trojan horses, spyware, keystroke loggers, etc.

Mashup: An application that uses elements from more than one source to create a new hybrid service. A well-known example is HousingMaps, a mashup of Google maps and housing information from Craigslist.

Member directory: A member directory displays information about community members in a roster format. It provides a broad overview of the membership, as well as some key information about each member. This information is either provided by members themselves or generated automatically; some may be required, and some may be optional. The contents may overlap with and link to an **individual profile** page. The **directory** may indicate the community's demographics or internal structure, such as distinct types of members, roles, or subgroups.

Metadata: Metadata is data about data, or information about information. For example, where a photograph is considered data, its metadata might include when and where it was taken, by whom, with what kind of camera, etc. When you play a music file in a media player and the names of the artist, song, album, and so on are displayed, that is metadata. Metadata provides context for the data and computer programs use it in a variety of ways. **Tags** are an example of metadata that are added incrementally over time.

Microblogging: Sending short posts such as those found in Twitter, which are limited to 160 characters. Microblogging lets users tell each other what they are doing throughout the day, in short bursts.

New indicators: Indicators that mark new items on a site, such as new postings in conversations, documents, member profiles, etc. In some systems, the number of such new items in a page and its subpages appears by the link to the page. The definition of "new" varies considerably, whether defined by recency (newly posted) or by whether you have seen the item (new to you).

Newsgroup: A public discussion forum found within the Usenet system (an early Internet bulletin board discussion system).

Newsletter: A regularly published document (electronic and/or paper) that is distributed to a community or subscriber group to disseminate key information.

Notepad: An electronic "scratch pad" that allows users of a system to keep notes or a journal about their activities in a system.

Online: As in online community, online education, online games, etc. Connected via computer and network. Contrasted with offline or **face-to-face**.

Open source software: "Generically, open source refers to a program in which the source code is available to the general public for use and/or modification from its original design free of charge, i.e., open. Open source code is typically created as a collaborative effort in which programmers improve upon the code and share the changes within the community. Open source sprouted in the technological community as a response to proprietary software owned by corporations." <u>Source</u>: Webpedia, http://www.webopedia. com/TERM/o/open_source.html

P2P (peer-to-peer): Refers to networks of computers that communicate directly with each other, rather than through a centralized server.

Participation statistics: Information that allows community leaders or members to monitor or visualize community participation on a **platform**. May include the ability to monitor activity down to the feature or location level, such as within a particular discussion.

Pattern language: A term originally used by architect Christopher Alexander to describe the designs of cities and buildings. Each pattern includes context or setting, problem, and solution. Pattern language has been applied to other fields or domains, from education to sustainability. Structured software design often uses pattern language. It has also been used to describe online communication, wiki use and roles, online communities, and more.

Permalink: "The URL of the full, individual article, designed to refer to a specific information item (often a news story or **blog** item) and to remain unchanged permanently, or at least for a lengthy period of time to prevent link rot." <u>Source</u>: Wikipedia, http://en.wikipedia.org/wiki/Permalinks

Personalization: Some tools allow users to change aspects of the interface to fit their own preferences. Examples include font size, time zones, language, layout, search order, and so on. Personal profiles are another kind of personalization.

Personal web pages: Web pages maintained by an individual. They may be part of a larger website, such as an online community, school, or department. See also **individual profile**.

Photo and video sharing: Websites where users can post and **tag** digital photos or videos and others can find and sometimes **rate** and comment on them. Some sites are free and others charge members to store and share. Examples of sharing sites include Flickr (photos) and YouTube (video). Digital photos, music, and videos can also be shared directly between personal computers with peer-to-peer (**P2P**) file sharing systems.

Platform: A technology package that integrates a number of tools.

Podcasting: A method of distributing audio files, putting the control of selecting, downloading, and listening in the hands of users to listen to on MP3 devices, allowing them to experience audio events across time and distance. It differs from a traditional broadcast model which determines what gets broadcast when. Some podcasts can be subscribed to and syndicated through an **aggregator** like **RSS**.

Polling: Creating and launching an online survey. A poll can include a variety of kinds of questions (single choice like Y/N, multiple choice, open-ended, etc.). Usually includes functionality to allow collection and analysis of collected data. Polling tools can operate in both synchronous and asynchronous environments.

Post: (*verb*) To contribute something to a group discussion, website, or other exchange, as in posting to an **email list**, a discussion forum, a **blog** or **wiki**, or a **photo or video sharing service**. (*noun*) The actual contribution (message, comment, file, etc.).

Presence indicators/tools: A presence indicator lets you know who else is active (on their respective computers) or logged on (to a system on which both parties are working). It is universally integrated with **instant messaging** and **chat** and is often useful in combination with other tools.

Question and answer tool: Allows participants to ask questions and receive answers. Unlike an email list, it manages the process quite explicitly by routing questions to "experts" and either routing answers back to the questioner or to the community.

Rating tools: Allow users to give numeric feedback. Used to quickly note the utility, correctness, or appropriateness of a posting or resource. Normally found in very large and anonymous communities. Sometimes used to rate a member's behavior or performance as part of a reputation system. See also **behavioral parameters**.

Registration: Creating a secure, private account with a user name and a password for an online interaction space in order to log in.

Remix: (*verb*) To take an existing piece of content (for example, a song or a video) and change it, sometimes by combining or juxtaposing the original material with new material. (*noun*) The resulting hybrid content.

Repositories (for documents or files): See **content management systems**.

RSS: Shorthand for **R**eally **S**imple **S**yndication. A method to allow users to subscribe to web-based content they want to read on a regular basis, delivered to an **aggregator** or RSS reader. Subscribing to an RSS feed saves users the time and effort of connecting to many different **blogs**, news sites, and so on to check for new **posts**. Instead, the RSS reader automatically checks for new content and displays it all in one place, either as summaries or in full, organized by subscription.

Scheduling utilities: Software tools for scheduling. See also **calendars**.

Screencast: A combination of an online slideshow, computer demonstration, or video and audio. "A screencast is a recording of computer screen output, usually containing audio narration typically published as a video file. Screencasts are typically created to produce software and web application demonstrations." <u>Source</u>: Wikipedia, http://en.wikipedia.org/wiki/Screencast

Search/search engines: Programs that perform searches on a single computer or across the Internet, such as Google.

Security management: All features for managing security for a platform or a tool that might be required. The advantage of centralizing many tools on one platform is that security tools and processes can be centralized and shared.

Shared calendars: Calendars that can be shared or edited by multiple users. A method for coordinating the schedules of a community or a group of people.

Single login system: A method to allow users to log in once and then have access to many different systems or resources.

Site navigation: Features that help members find their way around a community space, its various spaces for interaction, repositories, and subgroups, as well as its tools and facilities.

Slide and video presentation: Tool to allow users to present sequences of slides or combinations of video and slides online or in a face-to-face setting.

Smart mobs: Groups that use technologies to organize and communicate quickly and efficiently, often in real time. "Mobile communication devices, peer to peer methods, and a computation-pervaded environment are making it possible for groups of people to organize collective actions on a scale never before possible -- smart mobs, for better and for worse." Source: *Smart Mobs: The Next Social Revolution*, by Howard Rheingold, http://www.smartmobs.com/book/book_toc.html

SMS: Shorthand for **Short Messaging System**. Text messaging for mobile phones. SMS messages are limited to 160 characters. Also called **texting**.

Smartphones: Mobile phones that have advanced features. Newer smartphones include a variety of computer tools, such as web browsers, email clients, calculators, file systems and storage, etc., as well supporting **texting** (**SMS**) and taking and sharing photos and videos. Also called PC phones and PDA (personal digital assistant) phones.

Social context: The context in which communication tools are used, that is, the resources and communities that can support your use of a platform.

Social media: A general term to describe activities that involve social interaction, technology, and user-generated content. "Social media can take many different forms, including Internet forums, message boards, weblogs, **wikis**, **podcasts**, pictures and video. Technologies include **blogs**, **picture-sharing**, vlogs, wall-postings, **email**, **instant messaging**, music-sharing, crowdsourcing, and **voice over IP**, to name a few." <u>Source</u>: Wikipedia, http://en.wikipedia.org/wiki/Social_media

Social network analysis: Methods for analyzing the relationships (ties) between people (nodes) in a network. These include such things as who communicates with whom through what channels or media, for what purposes, who knows whom, who is related to whom (kinship), who links to whom, etc. Social network analysis can measure how dense a network is (numbers of ties), how central specific people are, who the intermediaries are, where the power blocks are, and so on. "Six degrees of separation" is a popular idea loosely based on social network analysis.

Social networking tools: Social networking tools provide network analysis and connection facilities to allow people to understand the community structure and influence it over time. They are often integrated with a variety of other tools including **calendars**, **chat**, **presence indicators**, group membership, connections finders, expert finders, and community entry pages. Social networking tools allow users to understand the roles of individuals in the community by providing information on how they relate to others. Not to be confused with social networking sites like MySpace and Facebook.

Social networking sites: Websites that offer social networking tools for individuals and groups, most often free and advertising supported. Examples include Facebook (http://www.facebook.com), Meebo (http://www.meebo.com), MySpace (http://www.myspace.com) and Ning (http://www.ning.com).

Social or shared bookmarking: See also **tags** and **folksonomy**. "In a social bookmarking system, users save links to web pages that they want to remember and/or share. These **bookmarks** are usually public, and can be saved privately, shared only with specified people or groups, shared only inside certain networks, or another combination of public and private domains. People who have access can usually view these bookmarks chronologically, by category or **tags**, or via a **search engine**. Most social bookmark services encourage users to organize their bookmarks with informal tags. They also enable viewing bookmarks associated with a chosen tag, and include information about the number of users who have bookmarked them. Many social bookmarking services provide web **feeds** for their lists of bookmarks, including lists organized by tags. This allows subscribers to become aware of new bookmarks as they are saved, shared, and tagged by other users." <u>Source</u>: Wikipedia, http://en.wikipedia.org/wiki/Social_bookmarking

Social or shared tagging: The practice of tagging resources collectively, using **social bookmarking** sites such as http://delicious.com.

Subscriptions: Subscription tools allow an individual to "subscribe" to some element, such as an online **newsletter**, daily **email** announcement, **blog feed**, or news **feed**. The person then gets some sort of alert to advise him of new information, or the new information is organized in some way for easier access, such as in an **RSS** reader.

Tags: Keywords or **category** labels users attach to items or elements, such as posts, webpage **bookmarks**, **wiki** pages, **blog** posts, photos, videos, and music in media sharing systems, etc. Tags help other people find similar items. **Social tagging**, collaborative tagging, and social indexing are other terms for creating a **folksonomy**, or user-generated taxonomy. Tag clouds are visual displays of tags used on a website, with the most frequently used shown in larger type. See also **social bookmarking**.

Telephony and teleconferencing: One-to-one, one-to-many, and many-to-many synchronous voice conversations. Technologies include individual phone calls, phone bridges, and conference call services, as well as web-based voice interaction tools, such as **VoIP** (Voice over IP). Skype is a well-known VoIP service.

Texting: Sending text messages between mobile phones, either one-to-one, one-to-many or many-to-many. See also **SMS**.

Threading: A thread is a topic in an online conversation in a forum, **discussion**, **chat**, or **email**. Threaded **posts** or messages are grouped together for easier reading. Online conversations are often multi-threaded. Threading allows users to read the topics or threads of interest.

Thumb tribe: People who are very skilled at using their thumbs for **texting**, using a small joystick, operating a **smartphone**, etc. "Another name for the thumb-proficient is the thumb generation. In Japan, they're called *oyayubizoku*, which means 'clan of the thumbs' or 'thumb tribe.'" <u>Source</u>: Word Spy, http://www.wordspy.com/words/thumbculture.asp

Trackback: See also **blog trackbacks**. "A Trackback is one of three types of linkbacks, methods for Web authors to request notification when somebody links to one of their documents. This enables authors to keep track of who is linking to, or referring to their articles. Some weblog software programs, such as Movable Type and Community Server, support automatic pingbacks where all the links in a published article can be pinged when the article is published. The term is used colloquially for any kind of linkback." <u>Source</u>: Wikipedia, http://en.wikipedia.org/wiki/TrackBack

Translation tools (automatic): Tools for translating text, including web pages, from one language to another. Also called machine translation. The results will be of lower quality than a native speaker or professional translator would produce, because the meanings of words depend on their context, which the computer cannot understand.

Upload: (*verb*) To copy and transfer a file from a user's hard drive to the web-based interaction space, thus making the file available to other members of the space. (*noun*) A file that is so transferred. The opposite of **download**.

Usenet: Established in 1980, the oldest computer network communications system still in use. Discussions in Usenet take place in newsgroups, which are topics similar to discussion forums. Users post messages to newsgroups and read them with newsreaders. There is no central Usenet server. Instead, newsgroup servers are hosted by many organizations and networked together so posts can be synchronized between them. Some Usenet newsgroups are moderated; that is, all posts go to a moderator who then approves them for general posting.

Video: "Video is the technology of electronically capturing, recording, processing, storing, transmitting, and reconstructing a sequence of still images representing scenes in motion." Source: Wikipedia, http://en.wikipedia.org/wiki/Video

Videoconferencing: Using video on both ends of a **distributed** meeting to allow two or more people to have real-time conversations with a view of the speaker and/or a view of all participants. Often combined with other synchronous tools, such as **teleconferences**, **application sharing**, or web meeting platforms. Video conferencing can range from a low-resolution add-on to **chats**, **instant messaging**, and similar tools, to high-capacity, TV-quality set-ups where people meet in special-purpose facilities. A point-to-point (two-person) video conferencing system works like a video telephone. Each participant has a video camera, a microphone, and speakers mounted on his or her computer.

Video feeds: RSS or other subscription **feeds** for **video**.

Vodcast: Also called vidcast. "Vodcast is an emerging term derived from the audio podcast and video." Source: Wikipedia, http://en.wikipedia.org/wiki/Vodcast

VoIP (Voice over IP): "Short for Voice over Internet Protocol, a category of hardware and software that enables people to use the Internet as the transmission medium for telephone calls by sending voice data in packets using IP rather than by traditional circuit transmissions of the PSTN [public switched telephone network]. One advantage of VoIP is that the telephone calls over the Internet do not incur a surcharge beyond what the user is paying for Internet access, much in the same way that the user doesn't pay for sending individual e-mails over the Internet." Source: Webopedia, http://practicallynetworked. webopedia.com/TERM/V/VoIP.html

Web 2.0: A term with many definitions, suggesting greater interactivity and participation than one-way static websites. Some say it is business finally embracing the web as a **platform**. Others suggest that its user-generated content will enhance collaboration and collective intelligence. "Web 2.0 is a term describing the trend in the use of World Wide Web technology and web design that aims to enhance creativity, information sharing, and, most notably, collaboration among users. These concepts have led to the development and evolution of web-based communities and hosted services, such as **social networking sites**, **wikis**, **blogs**, and **folksonomies**." Source: Wikipedia, http://en.wikipedia.org/wiki/Web_2.0

Web meeting tools: Also known as web conferencing, live meetings or web presentation tools, web meeting tools are usually a platform for synchronous online group interactions. These tools might include **VoIP**, **chat** rooms, participant lists, hand raising, **polling**, a shared **whiteboard**, **slide presentations** and **application sharing**.

Web tours: Lists of URLs that take you on a tour of one or more websites, usually organized sequentially or by topic, to try out and see what they're like. Sometimes these are captured in a screencast or video.

Whiteboard (electronic): An online drawing tool allowing multiple members to draw together simultaneously.

Widget: A mini-application that you can put on your desktop, webpage, social networking page, etc. to enhance your online experience or that of your visitors. Examples include everything from animals to weather. There are thousands of free widgets available. Also called gadgets. In programming, widgets are graphical user interface elements, such as buttons, dialogue boxes, sliders, menus, etc.

Wi-Fi: Shorthand for "**wireless fidelity**". Wireless networking technology. Wi-Fi allows users to connect their computers, mobile phones, game consoles, and the like to the Internet through a wireless access point, or hotspot. Wi-Fi is widely available through public hotspots, and millions of homes, businesses, and university campuses.

Wiki: "A wiki is a collection of web pages designed to enable anyone who accesses it to contribute or modify content, using a simplified markup language. Wikis are often used to create collaborative websites and to power community websites. For example, the collaborative encyclopedia Wikipedia is one of the best-known wikis. Wikis are used in businesses to provide affordable and effective intranets and for knowledge management. Ward Cunningham, developer of the first wiki software, WikiWikiWeb, originally described it as 'the simplest online database that could possibly work.'" Source: Wikipedia, http://en.wikipedia.org/wiki/Wiki *Wiki* is a Hawaiian word for quick or fast.

Yellow pages: An extended directory of people where they advertise what they know and how they could help others in the community.

Additional online material

We collect and make available additional material online at the *Digital Habitats: stewarding technology for communities* website: http://technologyforcommunities.com

Glossary Sources

A-Z of (nontech) networking, http://socialmedia.wikispaces.com/A-Z+of+%28nontech%29+networking

Full Circle Online Interaction Glossary, http://www.fullcirc.com/community/interactionterms.html

How to Talk Web, http://www.cdf.org/issue_journal/how_to_talk_web-2.html

Illinois Online Network Glossary, http://www.ion.uillinois.edu/resources/tutorials/communication/glossary.asp

Key terms in social media and social networking, http://socialmedia.wikispaces.com/A-Z+of+social+media

NetLingo, http://www.netlingo.com/index.php

Web 2.0 glossary, http://www.socialsignal.com/blog/alexandra-samuel/web-2-0-glossary

Webopedia, http://webopedia.com

Whatis, http://whatis.com

Wikipedia, http://wikipedia.com

Index

LearningTimes Community, 50
LeFever, Lee, 76
Leiner, Barry, 18
Lenzo, Amy, 24
libraries, 81. *See also* content, repositories.
lifecycle (of communities), **103-104**, 138, *150, 168*
listservs, 4-5, 15
literacy, xviii, 183-185
login, single login system, 38, 51, *90, 98*, 109-110, 212g
    example, 50
look and feel (of a site), 48, 52, 208g
lurkers, 9, 11, 77-78, 187, 208g
    See also, peripheral participation.

management tools, *80*
    change management, 136, 165-166
    project management, 79-81, 136
    site management, 61
    See also content management, knowledge management.
maps, see CommunityMap, geomapping.
March of Dimes community, 194, 196
Maron, Anna, 65
Maron, Mikel, 65
mashups, **66-67**, 127, 128, 208g
mass collaboration, 66, 174, **176**
    See also, wiki.
meetings
    orientation to, in communities, 70, *71*, **72-75**, *117, 153*
    tools for, *60, 62, 66*, 216g
    types, 72
members
    directory, *60*, **63**, *74, 85*, **88**, 204g, 208
    new, 139, 140, *167*
    profiles, *60*, **63**, 88, 207g
mentoring, 91. *See also* apprenticeship.
metadata, 209g, *see also* content management; tags.
Michalski, Jerry, 73
microblogging, *60, 74, 77, 85, 88, 95*, 209g. *See also* Twitter.
migrating communities (to a new platform), 135-138, *165*
military-related communities, 94
monitoring tools, 141-142, 145
Mosaic, 18
motivation
    forming communities, 4, 178
    stewardship, 28-29
MPD-SUPPORT-L, 4-9, 5n, 190
multi-topic conversations systems, 76, *77*
multimembership, 91, 92-93, 105-106, 178-179
multiplicity (of places for participation), 188

navigation (of a site), 24, 42, 44, 51-52, *90*, 212g
networking, 86-87, *88*
    mindset, 51
    stewardship role, 132-133
    See also relationship orientation; social media; social networking sites.
networks
    communities and, 12, 31, 84, **192-197**
    network-community hybrids, 196-197
    See also social networks, Usenet.
new communities, xix-xx, 104, **188-191**. *See also* development; lifecycle.
new indicators, 44, *60*, 209g
new members, 139, 140, *167*
new types of communities, **192-199**. *See also* proto-communities.
newsgroups, 15, 209g
newsletters, *60*, 63, 209g
non-profit communities, 24, 46, 65, 78, 127, 192, 194
notepad, *90*, 209g
NPtech, 192
NTEN network, 66

offline reading, 44, *43*
online access, 108, *151*
online communities, 209g
    development, 7, 103-104, 149, 173-174, *174*
    geographies, 177-181, 190-191
    history, 13-21
    new communities, 64, 104, 188-191
    new notions of community, **192-199**
    online place, 31, 50, 189
    outside connections, 105, *150, 157*, 185, 187, 197
    technical requirements for members, 108, 109, *151*
    variety of, 17
    See also communities of practice; digital habitats; formation of communities; lifecycle; orientations; technology landscape; *communities by name.*
online material for *Digital Habitats*, xx, xxvi, 148
online place for communities, 31, 50, 189. *See also* domain; homesteading.
open-source
    software, **125-127**, *161*, 209g
    communities, 19, 125-126, *161*
open-ended conversations
    orientation to, **75-78**, *77, 153*, 189-190
    examples, 4-10, 76, 118-119
O'Reilly, Tim, 17
organizations
    as context, 97, 105
    stewardship in, 30-32
    See also business communities.

Authors

Etienne Wenger (http://www.wenger.com)

Etienne Wenger is a global thought leader in the field of communities of practice and social learning systems. He is the author and co-author of seminal books on communities of practice, including Situated Learning, where the term was coined, Communities of Practice: learning, meaning, and identity, where he lays out a theory of learning based on the concept, and Cultivating Communities of Practice, addressed to practitioners in organizations who want to base their knowledge strategy on communities of practice. Etienne helps organizations in all sectors apply these ideas through consulting, public speaking, teaching, and research.

Nancy White, Full Circle Associates (http://www.fullcirc.com)

Nancy brings over 25 years of communications, technology and leadership skills in her work supporting collaboration, learning and communications in the NGO, non profit and business sectors. Grounded in community leadership and recognized expertise in online communities and networks, Nancy works with people to leverage their strengths and assets towards tangible goals and meaningful process. Nancy's blog and Twitter stream are regularly recognized as leading sources on online communities and networks, knowledge management and knowledge sharing. Nancy is a respected speaker and workshop leader. She is a chocoholic and lives with her family in Seattle, Washington, USA.

John D. Smith, Learning Alliances (http://learningalliances.net)

John Smith brings over 25 years of experience to bear on the technology and learning problems faced by communities, their leaders and their sponsors. He coaches and consults on issues ranging from event design and community facilitation, to community design and evaluation, and technology selection and configuration. He has been focused on communities of practice for the past 10 years and is the community steward for CPsquare, the international community of practice on communities of practice. He is a regular workshop leader in CPsquare and elsewhere. He grew up in Humacao, Puerto Rico and now lives in Portland, Oregon.

CPSIA information can be obtained at www.ICGtesting.com
Printed in the USA
238408LV00002B/49/P